Cityports, Coastal Zones and Regional Change

Cityports, Coastal Zones and Regional Change

INTERNATIONAL PERSPECTIVES ON PLANNING AND MANAGEMENT

Edited by

BRIAN HOYLE
University of Southampton

Published in association with the
Transport Geography Research Group
of the
Royal Geographical Society with the Institute of British Geographers
and the
Associazione dei Geografi Italiani

JOHN WILEY & SONS
Chichester · New York · Brisbane · Toronto · Singapore

Other Wiley Editorial Offices

John Wiley & Sons, Inc., 605 Third Avenue,
New York, NY 10158-0012, USA

Jacaranda Wiley Ltd, 33 Park Road, Milton,
Queensland 4064, Australia

John Wiley & Sons (Canada) Ltd, 22 Worcester Road,
Rexdale, Ontario M9W 1L1, Canada

John Wiley & Sons (Asia) Pte Ltd, 2 Clementi Loop #02-01,
Jin Xing Distripark, Singapore 0512

Library of Congress Cataloging-in-Publication Data

Cityports, coastal zones, and regional change : international
    perspectives on planning and management / edited by Brian Hoyle.
        p.   cm.
        'On behalf of the Transport Study Group of the Royal Geographical
Society with the Institute of British Geographers and the
Associazione dei Geografi Italiani.'
        Includes bibliographical references and index.
        ISBN 0-471-96277-5 (ppc)
        1. Coastal zone management.   2. Harbors.   3. Regional planning.
I. Hoyle, B.S.   II. Royal Geographical Society (Great Britain).
Transport Geography Study Group.   III. Institute of British
Geographers.   IV. Associazione dei Geografi Italiani.
HT391.C485   1996
338.91'7 – dc20   96-6467
                                                                 CIP

British Library Cataloguing in Publication Data

A catalogue record for this book is available from the British Library

ISBN 0 471 962 775

Typeset in 10/12pt Sabon by MHL Typesetting Ltd, Coventry
Printed and bound in Great Britain by Bookcraft (Bath) Ltd
This book is printed on acid-free paper responsibly manufactured from sustainable forestation, for which at least two trees are planted for each one used for paper production.

In memory of
Professor Giuseppe Rizzo
Faculty of Economics, University of Verona
1935–1996
who greatly stimulated the discussions on which this book is based
but whose untimely death occurred before its publication

Contents

Notes on Contributors   ix

Preface and Acknowledgements   xv

1   Ports, Cities and Coastal Zones: Competition and Change in a
    Multimodal Environment   1
    Brian Hoyle

Part I   THEORY AND PRACTICE IN THE UK   7

2   Port–City Relations and Coastal Zone Management in the Severn
    Estuary: The View from Bristol   9
    Keith Bassett and Tony Hoare

3   Managing the Cityport/Coastal Zone Interface: A Mersey Estuary
    Case Study   27
    David Massey

4   Cityport Development and Regional Change: Lessons from the
    Clyde   49
    Andrew H. Dawson

Part II   ENVIRONMENTAL ISSUES   59

5   Environmental Perception and Planning: The Case of Plymouth's
    Waterfront   61
    Luciano Cau

6   Oil Industry Restructuring and its Environmental Consequences in
    the Coastal Zone   83
    Sarah Harcombe and David Pinder

Part III   ECONOMIC DIVERSIFICATION IN DEVELOPING AREAS   103

7   Balkan Transport and Cityport Development in an Era of
    Uncertainty   105
    Derek Hall

8   Patras and its Hinterland: Cityport Development and Regional
     Change in 19th-Century Greece   121
     Elena Frangakis-Syrett and Malcolm Wagstaff

9   Cityport Development and Cultural Heritage: The Case of
     Thessaloniki, Greece   137
     Stella Kostopoulou

10   Cityports and Coastal Zones in Contemporary Africa: Mombasa
     and the Indian Ocean Façade of Kenya   159
     Brian Hoyle

11   Diversifying the Cityport and Coastal Zone Economy: The Role of
     Tourism   181
     Andrew Church

Part IV   INTERMODALISM, MIDAs AND MULTIMODALISM   211

12   Fixed Links and Short Sea Crossings   213
     Richard Knowles

13   The Venice Port and Industrial Area in a Context of Regional
     Change   235
     Stefano Soriani

14   Combined Transport in Italy: The Case of the Quadrante Europa,
     Verona   249
     Claudia Robiglio

Part V   PLANNING STRATEGIES AND IMPLICATIONS FOR THE
         FUTURE   271

15   Lessons from Massachusetts Coastal Zone Management's
     Designated Port Area Program: The Fore River Shipyard Re-use
     Project   273
     Laurel Rafferty

16   Cityports, Coastal Zones and Sustainable Development   295
     Adalberto Vallega

     Index   307

Notes on Contributors

Keith Bassett is a Lecturer in Geography at the University of Bristol. He studied for his first degree in geography at Birmingham University, UK, and after two years as a postgraduate at Pennsylvania State University, USA, returned to the UK to complete a PhD at Bristol. Dr Bassett's published work includes articles on urban politics, local economic development strategies and urban cultural policies. He was a member of the Bristol City Council in the 1980s, when some of the events discussed in Chapter 2 took place. He has previously cooperated with Tony Hoare on studies of port policy and local politics.
University of Bristol, University Road, Bristol BS8 1SS, UK

Luciano Cau graduated in geography from the University of Cagliari, Italy, and subsequently took an MSc degree at the University of Plymouth, UK. He is currently working towards a PhD degree in the Department of Geography, University of Southampton. His research interests include port cities and the geography of tourism.
University of Southampton, Southampton SO17 1BJ, UK

Andrew Church is a Lecturer in Geography at Birkbeck College, University of London. He lectures on urban and economic geography with a focus on Western Europe. His research interests include local economic policy, urban regeneration and the economic geography of hospitality and tourism. His current research projects include a study of tourism in London and the monitoring and evaluation of two City Challenge regeneration projects in Britain.
Birkbeck College, University of London, 7–15 Gresse Street, London W1P 1PA, UK

Andrew Dawson is a Senior Lecturer in the Department of Geography, University of St Andrews. As an economic geographer Dr Dawson has a longstanding interest in the management of the space economy (involving research in Poland, Japan and the USA) and in Scotland's estuarial regions. His publications consider, among other matters, the changing patterns of land use, regional development, transport and local government in Scotland since the Second World War. He has also recently published *A Geography of European Integration* (London: Belhaven, 1993).
University of St Andrews, Fife KY16 9ST, UK

Elena Frangakis-Syrett is an Associate Professor of History at Queens College and the Graduate Center of the City University of New York. Her major research interests are the economic history of the eastern Mediterranean, especially the Peloponnese and western Anatolia, during the 18th, 19th and early 20th centuries. Her major publications include *The commerce of Smyrna in the eighteenth century (1700–1820)* (Athens, 1992).
Queens College and the Graduate Center of the City University of New York, 65–30 Kissena Boulevard, Flushing, New York 11367–1597, USA

Derek Hall is Head of the Department of Leisure and Tourism Management at The Scottish Agricultural College. His research interests include tourism and transport development processes in restructuring societies and peripheral regions. His recent publications have included *Transport and economic development in the new Central and Eastern Europe* (editor, Belhaven, 1993), *Albania and the Albanians* (Pinter, 1994), *Tourism: a gender analysis* (co-editor with Vivian Kinnaird, Wiley, 1994), and *Reconstructing the Balkans* (co-editor with Darrick Danta, Wiley, 1996). Dr Hall has served as Secretary (1990–93) and Chairman (1993–96) of the Transport Geography Study Group of the Institute of British Geographers.
Department of Leisure and Tourism Management, The Scottish Agricultural College, Auchincruive, Ayr KA6 5HW, UK

Sarah Harcombe graduated in Environmental Sciences and is a postgraduate researcher in the Department of Geographical Sciences, University of Plymouth. Her research focuses on the environmental, social and economic consequences of industrial restructuring in the coastal zone, taking as its empirical focus the impacts of oil refining restructuring. In addition to her academic interest in coastal zone management, Ms Harcombe is an active member of environmental groups in South West England.
University of Plymouth, Plymouth PL4 8AA, UK

Tony Hoare studied for his first and graduate degrees in geography at the University of Cambridge, where his doctoral thesis explored the inter-relationships between air transport and industrial location. Transport and economic development have been important themes in his research and teaching subsequently, with particular reference to advanced economies. Between 1970 and 1976 he was a Lecturer in Geography at Queen's University Belfast, and subsequently became a Senior Lecturer in Geography at the University of Bristol. Dr Hoare is the author of *The location of industry in Britain* (Cambridge University Press, 1983) and co-editor of *Diffusing geography: essays for Peter Haggett* (Blackwell, 1995). He has contributed a number of papers and chapters to publications on several aspects of economic geography in Britain and also in New Zealand where he was visiting lecturer at the University of Canterbury in 1985. His present work with Keith Bassett is part of a long sequence of joint

research into a variety of aspects of Britain's trade and port development, and their economic and environmental implications, some of which use the Bristol context as illustration.
University of Bristol, University Road, Bristol BS8 1SS, UK

Brian Hoyle is a Reader in Geography in the Department of Geography, University of Southampton. His research interests include ports and port cities; waterfront revitalisation; and transport/development relationships, particularly in Africa and Canada. Recent publications include *Transport and development in tropical Africa* (John Murray, 1988); *Revitalising the waterfront: international dimensions of dockland redevelopment* (edited with David Pinder and Sohail Husain, Belhaven, 1988); *Modern transport geography* (edited with Richard Knowles, Belhaven, 1992); *European port cities in transition* (edited with David Pinder, Belhaven, 1992); and numerous papers on port development and waterfront change. Dr Hoyle served as Secretary (1987–90) and Chairman (1990–93) of the Transport Geography Study Group of the Institute of British Geographers, and is a Fellow of the Chartered Institute of Transport.
University of Southampton, Southampton SO17 1BJ, UK

Stella Kostopoulou is a Research Assistant in the Department of Economics, Aristotelian University of Thessaloniki, where she took her bachelor's and master's degrees in Economics and where she is currently working towards a PhD degree. She has also studied at the Universities of Grenoble and London (LSE). Her numerous publications consider regional development, transport planning, port cities and Balkan and Middle Eastern issues.
Department of Economics, Faculty of Law and Economics, Aristotelian University of Thessaloniki, 54006 Thessaloniki, Greece

Richard Knowles is a Senior Lecturer in the Department of Geography, University of Salford. Dr Knowles has served as Chairman (1987–90) of the Transport Geography Study Group of the Institute of British Geographers, and is Editor of the *Journal of Transport Geography*. In 1989 he undertook for British Rail, with John Farrington and Richard Gibb, the Channel Tunnel Act 1987 Section 40 Report on the Process of Regional Consultation. He has written widely on transport and edited *Modern Transport Geography* (Belhaven, 1992) with Brian Hoyle.
University of Salford, Salford M5 4WT, UK

David Massey lectures in urban studies in the Department of Civic Design, University of Liverpool, where he has been associated (in one capacity or another) with editing *Town Planning Review* since 1971. He was the inaugural Secretary-General of the Association of European Schools of Planning (AESOP) and is Secretary-Treasurer of the International Planning History Society. David Massey was a member of the University Study Team (1992–95) commissioned to prepare the Mersey Estuary Management Plan, covering topics related to land ownership, commercial navigation and port development and urban regeneration. His current research interests include coastal planning and management and environmental planning history.

Department of Civic Design, University of Liverpool, Abercromby Square, Liverpool L69 3BX, UK

David Pinder is Professor of Economic Geography in the Department of Geographical Sciences, University of Plymouth. His research interests integrate port studies, waterfront revitalisation and energy issues, particularly in a European context. Professor Pinder's publications on these themes include *Revitalising the waterfront: international dimensions of dockland redevelopment* (edited in 1988 with Brian Hoyle and Sohail Husain); *European port cities in transition* (edited in 1992 with Brian Hoyle); and numerous papers on waterfront change, oil industry restructuring and the European port and energy systems.

University of Plymouth, Plymouth PL4 8AA, UK

Claudia Robiglio is an Associate Professor in Human Geography in the Institute of Economic Geography, Faculty of Economics at the University of Verona. Professor Robiglio's major research interests include urban and transport geography and tourism.

Faculty of Economics, University of Verona, Via dell'Artigliere 19, 37129 Verona, Italy

Laurel Rafferty has had 17 years of urban planning and management experience in both the public and private sector. She holds the position of Harbour Planning Coordinator with the Massachusetts Coastal Zone Management (MCZM) Office with responsibility for management of the state's port and harbour revitalisation programme. Representing the state Executive Office of Environmental Affairs, she has served on the executive staff of both the Governor's Commission on Commonwealth Port Development and the Massachusetts Seaport Advisory Council. She acts in liaison with international port associations. In the private sector, she was a principal in Connery Associates, Inc., a land-use planning and development firm, and, in 1989, was President of the Massachusetts Association of Consulting Planners. She holds an MA from Tufts University and a BA from Wheaton College (Norton, MA).

Massachusetts Coastal Zone Management Office, 100 Cambridge Street, Boston, Massachusetts 02202, USA

Stefano Soriani is a researcher in the Department of Economics, University of Venice, and a PhD student in the Department of Urban and Regional Economics, Erasmus University, Rotterdam. His research interests include cityport development, environment and economic development in coastal areas, and

attitudes towards the environment in Italy.
Department of Economics, University of Venice, Dorsoduro 3246, 30123 Venice, Italy and Department of Urban and Regional Economics, Erasmus University, Burgemeester Oudlaan 50, 3000 DR Rotterdam, The Netherlands

Adalberto Vallega is Professor of Urban and Regional Geography, Faculty of Architecture, and Head of POLIS, Department of Urban, Regional and Landscape Planning at the University of Genova. Founder and former President of the International Centre for Coastal and Ocean Policy Studies (ICCOPS), a United Nations Non-Governmental Organization based in Genoa, he was from 1988 to 1992 Chairman of the Commission on Marine Geography, International Geographical Union. He has published several volumes and numerous papers on maritime subjects, including *Per una geografia del mare* (Milan: Mursia, 1982), *Ecumene oceano* (Milan: Mursia, 1985), *Changing waterfront in coastal area management* (Milan: Franco Angeli, 1992), *Sea management: a theoretical approach* (London: Elsevier) and *Governo del mare e sviluppo sostenibile* (Milan: Mursia).
Facoltà di Architettura, Istituto di Urbanistica, Università di Genova, Stradone de S. Agostino 37, 16123 Genova, Italy

Malcolm Wagstaff is a Professor of Geography in the University of Southampton. His research interests include settlement and population change in the eastern Mediterranean, with special reference to southern Greece and Anatolia. He is currently working on the Peloponnese in the early 18th century. His major publications include: *The evolution of Middle Eastern landscapes: an outline to AD 1840* (London, 1985) and, as editor and contributor, *Landscape and culture: geographical and archaeological perspectives* (Oxford, 1987).
University of Southampton, Southampton SO17 1BJ, UK

Preface and Acknowledgements

Relationships between ports, cities and coastal zones are matters of widespread interest and concern, given the high and rising proportion of the world's population living in such locations and the increasing pressures upon such environments. Changing attitudes towards port–city relations, towards the role of port cities in littoral development, and towards the wider environmental and ecological issues involved in coastal zone management, have given rise in recent years to increased awareness of the problems involved and to a substantial literature. This book, the origins of which lie in an international conference held in Venice in March 1994, is designed to contribute to these debates on the basis of a wide range of contrasted but complementary experience.

The Transport Geography Study Group of the Institute of British Geographers first held a British–Italian Seminar in Transport Geography in 1989 at Royal Holloway and Bedford New College, University of London. On that occasion, the focus was primarily on urban transport themes and Italian participants came primarily from Rome. For some time members of the Group had felt that a second such seminar, to be held this time in Italy, would be useful and appropriate, particularly in view of the growing strength of contacts with Italian geographers working in the related fields of port-city development and coastal zone management. With the financial support of the Institute of British Geographers, the British Council and the Italian Government Research Council, the second seminar took place between 11 and 13 March 1994 in Venice.

The theme of the seminar was 'Cityport development and regional change', and was designed to build on recent work on the geography of ports and the port/city interface by directing attention specifically to interlinkages between port-city changes on the one hand and the problems of coastal zone management on the other, a field in which much useful work has been done by Italian geographers. The seminar was jointly convened by Dr Brian Hoyle (University of Southampton), on behalf of the Transport Geography Study Group of the IBG, and by Professor Gabriele Zanetto (Venice), most ably assisted by Dr Stefano Soriani (Venice), on behalf of the Associazione dei Geografi Italiani; and was held in the Istituto Salesiano on the Isola di San Georgio Maggiore, a perfect setting for a small-group seminar. Participants came from a variety of European and other countries, and papers were presented on contextual and methodological issues, on recent and current research, and on policies and planning experience. A full report on the seminar appeared in the *Journal of Transport Geography* 2(3), 211–13.

Selected papers presented at the seminar have been extensively revised and edited for the present volume, the structure of which is outlined in an introductory chapter. Editorial policy has been to work closely with authors whose contributions were seen particularly to illuminate, within a balanced structure, contrasted policies and planning issues.

I am grateful to my fellow contributors for their cooperation in the preparation of this book, and to Dr Iain Stevenson and his colleagues at John Wiley & Sons Ltd for their support and encouragement. Some of the maps and diagrams were prepared or amended in the Cartographic Unit of the University of Southampton, under the direction of Alan Burn. Others were submitted by authors in a finished condition, thus accounting for some variation in cartographic style. I also wish to record my appreciation of secretarial help provided by Jackie Bailey of the University of Southampton's Department of Geography.

Dr Brian Hoyle
University of Southampton

1 Ports, Cities and Coastal Zones: Competition and Change in a Multimodal Environment

BRIAN HOYLE

Department of Geography, University of Southampton, UK

In the advanced countries of the modern world, and in the developing countries too, in the past and today, cities and ports are frequently, indeed normally, intertwined in their location, development, functions and problems. At various scales and in a range of economic contexts, a port acts as a gateway and as a node within a series of transport networks, while a city is essentially a central place within a series of socio-economic and political systems. Gateway functions and central place functions are not invariably compatible. Nevertheless, the port function, wherever it has been developed, has normally given rise to some degree of urban growth, so that port–city relationships have become complex and well-established. The relationship between city size and port throughput – both dependent upon a wide range of factors including the specific character of port and urban activities in particular locations – has, however, always been highly variable and indeed somewhat tenuous; and, from the 1960s onwards, the evolution of maritime technology (involving particularly the widespread development of bulk terminals, container ports and roll-on/roll-off methods of cargo handling) has weakened the traditionally strong functional ties between ports and cities (Hoyle and Hilling, 1984).

Relationships between cityports and regions are also clearly recognised as important and varied. The interfaces between the cityport and the regions within which it is located and those which it serves introduce different sets of relationships. These may be less well-defined than in the case of the port/city interface, although they may reflect a similar variety of contexts – environmental, economic, political, etc. – and a similar range of issues such as transport, employment and planning. A significant difference arises, however, in terms of the spatial scales involved; for whereas the port/city interface is largely confined within the relatively discrete context of the built-up area, the cityport/region interface is not only far more extensive geographically but also involves in other

Cityports, Coastal Zones and Regional Change. Edited by Brian Hoyle.
© 1996 John Wiley & Sons Ltd.

dimensions a variety of different concepts, scales and levels of interaction, many of which are explored in this book.

There is also an important difference between the links between a port and, on the one hand, the complex set of hinterlands it serves (on a local, national and international basis) as well as a great variety of forelands across the oceans; and, on the other hand, the links between a cityport and the coastal zone within which it is located. These two sets of links clearly overlap, but the port–hinterland links involve functional connections through the medium of specific transport modes, while the cityport–coastal zone links are more likely to involve a more varied symbiosis. The interdependence of cityport and local region, therefore, involves only a limited proportion of the elements and functions characterising a cityport economy, for a port's *raison d'être* is to serve its hinterlands in their entirety.

Recent decades have witnessed substantial changes in port–city and cityport–region relationships. In the advanced world, and increasingly in developing countries too, the migration of port activities towards deeper water, as a consequence of technological change, has become an increasingly common phenomenon. This has introduced in many ports around the world an unaccustomed separation of port and urban functions, and the consequent redevelopment of older port zones and inner city areas. Similarly, the role of the cityport in a regional development context is beginning to receive increased attention, in terms of the impacts of rapid urban growth and port activity on neighbouring coastal environments, and in terms of the search for a more balanced, integrated approach to the management of port–city regions.

These issues and relationships have been complicated in recent years, both in advanced and in developing countries, by the introduction of new inland intermodal transport nodes, variously described or categorised as inland container depots (ICDs), freightliner terminals or freight villages. This phenomenon is essentially an outcome and extension of the growing trends towards intermodalism and multimodalism in international transport and, at least in advanced countries, but also increasingly in developing countries too, the flexibility of the new transport systems is having the effect of reducing the relative significance of traditional ports within the overall transport complex of a country or region. The modern port is no longer simply a terminal or break-of-bulk point, no longer simply a point where the mode of transportation changes from land- to water-borne systems. In the brave new world of intermodal and multimodal through transport, there are many such nodes, terminals and points, and the traditional cityport must necessarily adjust to changed and changing circumstances in terms of its competitive position and its planning context.

These trends serve to enhance rather than to reduce the interest associated with port–city and cityport–region relationships. The changing port–city relationship has given rise to numerous studies of the redeveloping waterfront, both thematic (Hoyle *et al.*, 1988) and idiographic (Breen and Rigby, 1993). The analysis of urban waterfronts and the transport problems of cities on water have been subjects of attention at two recent conferences organised by the Centro

Internazionale Città d'Acqua, Venice, Italy (Bruttomesso, 1993, 1995). A parallel but hitherto less well-developed research theme concerns cityports in their regional context, notably within the framework of coastal zone management and its extensive environmental/ecological literature, but also more widely within the regional planning structures on various scales and of many types within which port cities continue to occupy a pivotal role. An early attempt to look beyond the port–city relationship towards cityport–region problems was made at the 1990 Annual Conference of the Institute of British Geographers, held in Glasgow, UK, where a productive session of the Transport Geography Study Group led to a publication on *Port cities in context* (Hoyle, 1990). Subsequently, the discussion has been taken further forward by Vallega (1992), from the viewpoint of cityports in a coastal zone context, building on earlier work by Da Pozzo *et al.* (1985) on coastal planning perspectives. On a wider international front, the Association Internationale Villes et Ports has produced, from its base in Le Havre, France, a wealth of documentation and has organised a series of biennial conferences. The 1993 conference (held in Montreal, Canada) took as its theme the relationships between port cities and the environment (AIVP, 1994), while the 1995 conference (held in Dakar, Sénégal) directed attention to cityport–hinterland relationships especially as viewed from the hinterland in a context of interport competition.

The present work, arising directly from British–Italian cooperation in the field of transport geography, is intended as a further contribution to this broad debate. The book is divided into five parts, the first of which (Chapters 2–4) is concerned with theory and practice in the UK. Three chapters provide useful comparisons and contrasts between the recent development problems and experience of the Severn, Mersey and Clyde estuaries. Keith Bassett and Tony Hoare, in their chapter on the changing relationships between port and city in the Severn Estuary, begin with a brief survey of port development in the estuary as a whole and a summary of the evolution of the port of Bristol. The main thrust of their chapter concerns, however, two themes derived from this context: the recent impacts of privatisation on port–city relations and inter-port competition in a context of continuing port overcapacity; and the relationship between ports and development pressures in the Severn Estuary and the possible effects of emerging, estuary-wide environmental agendas and calls for integrated coastal zone management strategies. In his comparable study of the Mersey Estuary, David Massey analyses the background to, and methodology of, an integrated approach to coastal redevelopment at a regional scale, and highlights some of the problems and objectives involved. In a not dissimilar vein, Andrew Dawson emphasises the unifying factor of the Clyde which provides an example of the changing relationship between economy, settlement and port development at the regional as well as at the urban scale. Together these three studies provide many useful insights into concepts and approaches concerning cityport/coastal zone interfaces and many constructive ideas showing how current and anticipated changes might be managed.

In Part 2 (Chapters 5 and 6) attention turns to more specific environmental issues. Here, three authors present analyses of specific problems which, although using British evidence, illustrate from an environmental perspective some complex perceptions and interdependencies involved in coastal zone management and in port–city development. Luciano Cau draws on his recent research on environmental perception and planning processes in the port city of Plymouth, using an explanatory model of the influence of environmental perceptions on development and recommending a more flexible multidisciplinary approach towards territorial planning. Sarah Harcombe and David Pinder, in contrast, discuss the negative economic effects and the positive environmental outcomes of declining coastal zone industries. Using an investigation of UK coastal oil refinery closures, they assess the delicate and complex balance between environmental gains and costs in this context.

The third part of the book (Chapters 7–11) groups together a series of chapters broadly concerned, in widely differing contexts, with problems of economic diversification in developing cityports and coastal zones. In the first of these, Derek Hall looks at some of the implications of Balkan conflict and tension for transport and cityport development in south-eastern Europe. He focuses on the plans being developed to free current bottlenecks in the circumnavigation of Serbia, and the emerging role of Albanian ports and landward transport links, with their very limited current capacities and inferior infrastructures. Two studies focused upon Greek port cities and their regions then follow. In the first of these, which is a study in historical geography, Elena Frangakis-Syrett and Malcolm Wagstaff outline relationships between port development and hinterland population growth in the case of Patras. Their focus is on the 19th century and on the link between the cityport and its hinterland provided by the production and export of currants: expansion in the currant trade was reflected in population growth in both the city and its hinterland, much of it due to immigration, but the collapse of the trade towards the end of the century brought widespread socio-economic distress which resulted in the transformation of Patras into the leading emigrant port of Greece.

In a second study of a Greek cityport, Stella Kostopoulou discusses Thessaloniki and its Balkan hinterland, outlining the role and significance of the port city in a regional context throughout its long history and emphasising the importance of urban (particularly waterfront) revitalisation as opposed to inappropriate renovation. Two further chapters within this third part of the book also link past experience with present-day problems, and look forward in different contexts to some reorientation in cityport–coastal zone relationships in formal and functional senses. Brian Hoyle uses Mombasa and the Indian Ocean façade of Kenya as an illustration of cityport–coastal zone relations in a part of the less-developed world. Andrew Church discusses the role of tourism in the diversification of cityport economies, using several examples (Hartlepool and Teesside, Dover, and the London Docklands) to illustrate different tourism strategies adopted.

In part 4 (Chapters 12–14) attention turns to issues involving intermodalism, multimodalism and maritime industrial development areas (MIDAs). In the first of three chapters in this section of the book, Richard Knowles examines the impact on cityports and coastal zones of fixed road/rail links which replace ferry routes, using the Eurotunnel link under the English Channel and other case studies from Japan, Scandinavia and Scotland to illustrate how fixed links can actually or potentially affect city–region–hinterland expansion and competition. In a second chapter, Stefano Soriani considers the radical changes throughout the Venice region produced by the polarising effects induced by the growth and development of Porto Marghera, and argues for the development of new positive inter-relationships between the component elements of the Venetian cityport and its wider region, involving a reformulation of the regional role of the port complex. A third contribution, authored by Claudia Robiglio, explores the concept of inland intermodal centres with special reference to freight villages in Italy, using the example of Verona to demonstrate substantial changes in urban land use and regional transport flows consequent upon the development of intermodal combined transport systems.

The two chapters in the fifth and final part of the book (Chapters 15 and 16) serve effectively as a conclusion to the volume. They address planning strategies and implications for the future and place all the preceding studies into broader contexts by introducing first a transatlantic dimension to the European debate and second a view of cityport–coastal zone inter-relationships in a context of sustainable development. Laurel Rafferty analyses, from her perspective as a practising coastal zone planner, some lessons derived from the experience of Massachusetts, USA, where the state's coastal zone management authority has implemented a redevelopment programme for a series of designated port areas. By focusing on one very specific case study within a broad context, this chapter provides an insight into perceptions and practices from which other practitioners and students of cityport and coastal zone change elsewhere might glean some useful lessons. Finally, Adalberto Vallega provides an overview of cityports and coastal zones in which he combines perspectives derived from geographical literature, from the urban waterfront regeneration movement and from experience of coastal zone management, and attempts to relate these diverse elements to the concept of sustainable development. This forward-looking synthesis demonstrates again the need for evaluation to focus on environmental, ecological and socio-economic aspects of change within cityports and coastal zones and on the basis of an interdisciplinary perspective to prepare research agendas for the 21st century.

It is appropriate to conclude this introductory chapter with an apposite quotation from earlier work on the themes which this book addresses. Adalberto Vallega, in his capacity as Chairman of the International Geographical Union's Commission on Marine Geography, has done most to stimulate and encourage an integrated approach to coastal area management of the oceans. In this context, he has written (Vallega, 1992, 116–17) that

Waterfront evolution is characterized by growing complexity ... (but) its current phase allows us to evaluate the role of the waterfront in the general context of the coastal economy. Two alternative scenarios emerge: on the one hand, a scenario characterised by the absence ... of interactions between the waterfront and coastal area planning and management; on the other hand, a scenario based on strong integration between these two elements ... In short, because of the strategy of decision-making centres, the role of the waterfront ranges between two extreme conditions: the waterfront as a foreign body and the waterfront as a leading component of the coastal area ... In order to prevent the underevaluation of the role and importance of waterfront plans ... it is necessary intensively to consider the relationships between what takes place at the waterfront ... and in the coastal area as a whole. The knowledge and awareness of these relationships may open the way to new horizons of coastal area planning and management.

In different ways this quotation and the chapters which follow outline and reflect a new reality of relationships and power distribution between ports, cities and coastal zones, and move towards a possible set of principles framing coastal zone management in a context of sustainable development for the future.

REFERENCES

AIVP (1994), *Villes portuaires, acteurs de l'environnement* (4th International Conference, Cities and Ports, Montreal, 1993) (Le Havre: Association Internationale Villes et Ports).

Breen, A. and Rigby, D. (eds) (1993), *Waterfronts: cities regain their edge* (New York: McGraw-Hill, for the Waterfront Center, Washington, DC).

Bruttomesso, R. (ed.) (1993), *Waterfronts: a new frontier for cities on water* (Venice: International Centre Cities on Water).

Bruttomesso, R. (ed.) (1995), *Cities on water and transport* (Venice: International Centre Cities on Water).

Da Pozzo, C., Fabbri, P. and Vallega, A. (1985), *Coastal planning: realities and perspectives* (Genoa: Municipality and University of Genoa, for the International Geographical Union).

Hoyle, B.S. (ed.) (1990), *Port cities in context: the impact of waterfront regeneration* (edited for the Transport Geography Study Group, Institute of British Geographers).

Hoyle, B.S. and Hilling, D. (eds) (1984), *Seaport systems and spatial change: technology, industry and development strategies* (Chichester: Wiley).

Hoyle, B.S., Pinder, D.A. and Husain, M.S. (eds) (1988), *Revitalising the waterfront: international dimensions of dockland redevelopment* (London: Belhaven Press).

Vallega, A. (1992), *The changing waterfront in coastal area management* (Milan: Franco Angeli).

Part I

THEORY AND PRACTICE IN THE UK

2 Port–City Relations and Coastal Zone Management in the Severn Estuary: The View from Bristol

KEITH BASSETT and TONY HOARE
Department of Geography, University of Bristol, UK

THE CONTEXT: THE PORTS OF THE SEVERN ESTUARY AND REGIONAL DEVELOPMENT

The Severn Estuary is well supplied with ports, which have had varied histories and different relationships to local economic development. Bristol, for example, was already a major port in the Middle Ages and the centre of a great trading empire, but the ports along the South Wales coast mainly owed their development to 19th-century industrialisation and the growth of the local coal industry. Many of them reached their peaks as coal-exporting ports early in the 20th century before entering a long period of relative decline.

By the 1960s a general pattern had been established. The Severn ports all tended to concentrate on handling bulk import cargoes such as ore, timber and feedstuffs, mainly for local industries, with only limited manufactured exports (Figure 2.1a) (Bird, 1963). Since 1980 the ports have experienced mixed fortunes as UK trade has shifted towards east coast ports, and by 1991 the combined ports were only handling around 28 millions tonnes of cargo or 8% of Britain's dry bulk trade (Figure 2.1b).

In the mid-1990s there is still considerable excess port capacity in the estuary given likely forecasts of bulk cargo trade, while competition between the ports has intensified as a result of recent privatisations. In this chapter we explore these developments in more detail by focusing on the case of Bristol, historically the most significant of the Severn ports.

BRISTOL: THE EVOLUTION OF PORT–CITY RELATIONS

The history of the port of Bristol may be described within a broad 'stages' model of development.

Cityports, Coastal Zones and Regional Change. Edited by Brian Hoyle.
© 1996 John Wiley & Sons Ltd.

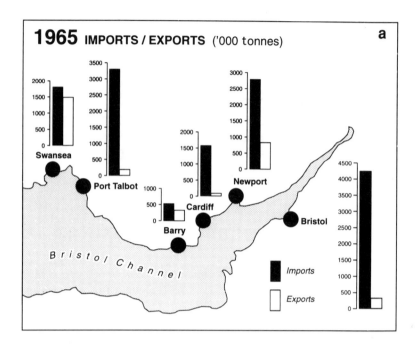

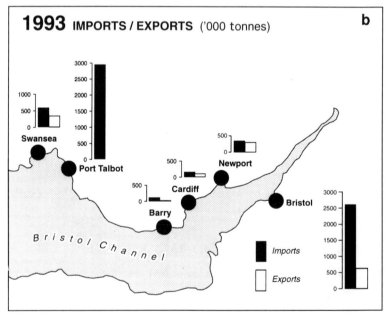

Figure 2.1.   Seaborne trade of ports in the Severn Estuary, 1965 and 1993

The medieval and early modern cityport

Bristol was one of the great ports of Europe during the medieval period. Indeed, from the 17th century until well into the 18th century Bristol was not only the second port in the country but also its second largest city, a hub of a great trading empire extending from West Africa to North America (Sacks, 1991; Morgan, 1993). During this period trade was centred on the tidal wharves in the heart of the city, three miles (5 km) up the Avon from its confluence with the Severn (Figure 2.2). However, by the end of the 18th century Bristol had lost ground to Liverpool and Glasgow, partly due to the difficulties of access to the port for

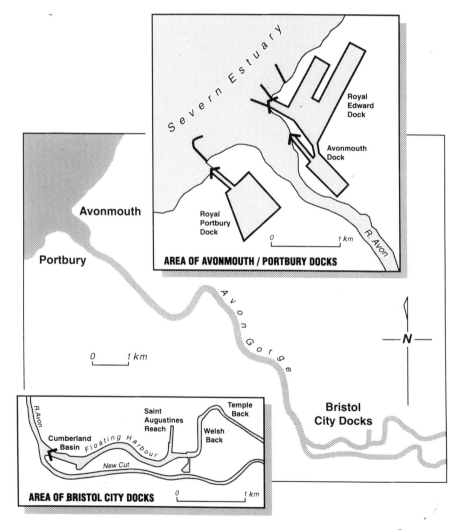

Figure 2.2. Historical evolution of port facilities in the Bristol region

larger ships which had to contend with the winding passage up the Avon Gorge and the second highest tidal range in the world. For the next 150 years the city struggled to restore its former maritime greatness through a series of major docks projects.

The 19th-century expanding cityport

In 1804 the city's merchant élite financed a major port improvement scheme which dammed and diverted the Avon to form a floating harbour in the heart of the city. However, the scheme ran into financial difficulties and soon had to be taken over by the City Council and run as a municipal enterprise. The still-increasing size of ships soon threatened to make these city-centre docks obsolete, and in 1877 new deep-water docks were opened downstream at the river's mouth at Avonmouth, capable of handling 30 000-ton vessels (Figure 2.2). These in turn were soon taken over by the Council and run as an integral municipal enterprise with the old city-centre docks.

The 20th-century maritime industrial cityport

The new docks became the centre of an important industrial sub-region after the First World War following the construction of a large lead and zinc smelter and a fertiliser plant, both supplied with raw materials through the port. Although the port expanded its range of bulk import cargoes over the succeeding decades, exports were much more limited, and the port never acted as a significant outlet for manufactured goods, local or national (Bird, 1963). Although the adjacent industrial zone was extended with a new chemical complex in the 1960s, this did not generate much extra port traffic. The docks also had to operate in the regional context of dock overcapacity in the Severn Estuary, competing with the numerous South Wales ports for similar kinds of bulk import cargoes.

It was in this context that in 1964 Bristol decided to expand dramatically its port system to handle a new generation of ever larger ships. This decision was fuelled by the persistent belief, held both by local politicians and shipping interests, that the future of the city was still tied to its role as a great port. The first plan, submitted in 1964, envisaged a huge new Severnside superport, but was refused government backing after a detailed investment appraisal indicated its lack of economic viability (Bassett and Hoare, 1984). However, intense lobbying against the plan by Labour Party MPs in South Wales also played an important part in blocking the scheme, itself indicative of the long-standing economic rivalry between the two sides of the estuary.

After a second proposal was also turned down, the City Council finally achieved success in 1971 with a Private Member's Bill in Parliament. The government acquiesced on the clear understanding that Bristol would finance the entire scheme itself with no government subsidy. Although the Council envisaged the new port as the centre of a new industrial zone, a government-funded study of the whole Severnside region, published in 1971, ominously concluded that the local ports would not be significant factors in the future growth of the region (DOE, 1971). Nevertheless, the city went ahead, borrowed the necessary capital, and the new docks opened in 1977 (see Figure 2.5). Bristol became the proud owner of the largest municipal docks in the country with the largest dock gates in the Severn Estuary. Unfortunately, the scene was set for disaster rather than triumph.

Bristol's new port opened at the start of a recession, but in the longer run it was affected more seriously by the shift of overseas trade towards east coast ports. As a result, between 1980 and 1990 Bristol slipped from 10th to 17th place in the British port hierarchy in terms of non-fuel trade. Figure 2.3 shows the trading and financial consequences. After the severe slump in the early 1980s overall trade in 1991 had climbed back to 1980 levels, but the number of port employees had been cut from over 2000 to 704. However, the annual deficit rose to almost £14 m in 1982, requiring a huge subsidy from the Bristol ratepayers of £10 m. Debt restructuring and a new development plan to improve infrastructure

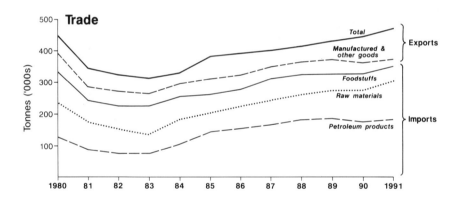

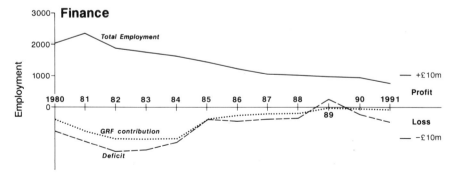

Figure 2.3. Trading and financial performance of the Port of Bristol, 1980–91 (GRF: General Rate Fund)

and labour productivity gave the docks some breathing space in the late 1980s, but the gathering force of a new recession and a national dock strike plunged the port into further losses. The Labour-run council had also been targeted as a 'loony left' council by the Conservative Government and it was hit hard by poll-tax capping. By 1990 the docks were a major financial burden on the city.

Closure was contemplated but rejected. The dock expansion had, after all, been supported by both the major parties on the council with few dissenting voices. There was also the question of jobs. In many ways the predictions in the 1971 Severnside report, that the future growth of the sub-region was not linked to the ports, proved correct. In the 1980s the Bristol region grew rapidly, acquiring the reputation of a 'sunbelt city' at the end of the 'M4 corridor' of high-technology growth (Boddy et al., 1986). In fact, although the local aerospace companies remained important, most of this growth was generated by the rapid expansion of the service sector, particularly in the field of insurance, banking and finance. The port thus played only a minor role in this growth pattern, and the heavy industrial area of Avonmouth proved unattractive to most of the new companies, which preferred to locate either in the city centre or in the new business parks on the northern fringe of the city next to motorway junctions (Bassett, 1986). Nevertheless, the Avonmouth area remained an important focus of employment, and a 1981 consultant's report estimated that around 6500 jobs would be lost in the port itself and in port-related industries if the port closed, raising the local unemployment rate to 25%

It is understandable, therefore, that the council decided to struggle on with its port, trimming the workforce where possible and trying to adopt a more aggressive and entrepreneurial management style. In this it was hampered by the ruling Labour group's traditional suspicion of business motives and its hostility towards job cuts and land sales at the port. However, there were wider changes at work that were dramatically reshaping the context of port competition.

PRIVATISATION AND REJUVENATION: A NEW STAGE OF PORT DEVELOPMENT?

Privatisation, which has affected all the major ports of the Severn Estuary, has presaged a new stage of port development. To explain how Bristol came to privatise its port we need to place local events in the wider national context.

Port privatisation: the national context

First, the 1980s saw a major change in the policy context. The Conservative Government elected in 1979 inherited a national port system with well-documented problems of overcapacity and inefficiency (Asteris, 1988; Turnbull and Weston, 1992). In line with its competitive, free-market orientation the government's solution was privatisation and deregulation to encourage greater competition. This strategy unfolded gradually (Bassett, 1993). Consequent upon the 1981 Transport Act and the privatisation of the British Transport Docks Board, 19 ports were sold off to form a new private company, Associated British Ports (ABP). By 1990 ABP had expanded to 22 ports, was making a profit of £60.2 m per year, and had diversified into land and property development (ABP, 1990).

Secondly came the abolition in 1989 of the National Dock Labour Scheme. The scheme, designed to end the use of casual labour on the docks, dated back to 1947 and had long been opposed by employers who argued that it encouraged overmanning and labour inflexibility. The abolition of the scheme, bitterly opposed by trade unions and by the Labour Party at the time, resulted in big increases in labour productivity and company profits, but at the expense of significant job losses (Turnbull, 1991).

Finally, the 1991 Ports Bill was directed at the privatisation of the trust ports. There were over 100 of these in 1991, some dating back to the 14th century, governed by varying and sometimes archaic constitutions. The larger trust ports were now required to bring forward plans for privatisation within two years. The bill was again opposed by Labour, and this time also by some Conservatives representing constituencies with successful trust ports, but was passed in 1991. A series of trust port privatisations followed (Clyde, Forth, Medway, Tilbury, Tees and Hartlepool) which in some cases proved highly controversial (*Financial Times*, 1992; Committee of Public Accounts, 1994). The most notorious case was that of Medway, sold to a private company for £29.7 m; it was resold 18 months later for £103 m after the sacking of much of the workforce.

The cumulative effect of this legislation was to significantly change the ownership structure of the British port industry and open up a new and more competitive era of port development. Although the new legislation did not directly affect Bristol, this new competitive environment was soon felt in the Severn and compounded Bristol's problems. When ABP was formed in 1981 it acquired ownership of all the major ports on the other side of the Severn in South Wales. The municipally controlled port of Bristol now found itself competing against a more aggressive private company. Not all the ports posed an equal threat, however. Port Talbot's trade was tied closely to the local steelworks. Swansea was too far down-channel to seriously threaten Bristol. Barry's trade was largely limited to banana imports. Cardiff's docks were incorporated within an Urban Development Corporation, which saw the future more in terms of property development for housing, office, and leisure use than in the expansion of port traffic. Newport, on the other hand, was only a few miles across the channel from Bristol, and imported similar kinds of bulk cargoes. Its trade expanded significantly in the early and mid-1980s.

The final piece of the jigsaw concerns the knock-on effects of changes in British energy policy, being transformed by the same privatisation strategy which affected British ports. The key factor here was the decision to privatise the electricity industry. Hitherto, the state-run generating utility had enjoyed a near monopoly of electricity provision but had also been locked into government-imposed contracts to buy inputs from the nationalised coal industry. Amid much controversy, privatisation now created a duopoly of two main generators, National Power and PowerGen. Nuclear power continued to be subsidised. The 12 newly privatised regional electricity suppliers were also allowed to take out 15-year contracts for gas-fired electricity from new companies they established and owned. In this much more competitive new environment the two privatised generators quickly saw the necessity for profitability of switching from the previous long-term contracts with British Coal to cheaper foreign imports to supply their coal-fired power stations. Imported steam coal for power stations rose from 1.3 million tonnes in 1989 to 7.3 million tonnes in 1991, prompting a search for new sites to construct bulk import terminals. Bristol emerged as a possible site, and port privatisation offered the means.

Privatisation at Bristol

Bristol had received an offer from ABP to buy its port in the late 1980s, but on terms the City Council found unacceptable. Then in 1990 the council received a further and more attractive offer from a company called First Corporate Shipping Ltd to purchase the port as a going concern (Bassett, 1993).

First Corporate was essentially its two directors, Terence Mordaunt and David Ord, whose background was in shipping and port management. According to the directors, they had been looking around for several years for a large port installation to run and early on had identified Bristol as a port with potential. The major advantages they perceived were as follows. First, Bristol was better located for reaching a large hinterland population than any of its major competitors. Over 42 million people lived within a day's lorry journey (250 km), compared with 33 million for Southampton and 31 million for Tilbury. Second, the port had the biggest lock in the country, capable of handling ships up to 120 000 tonnes deadweight. Third, the port had the advantage of direct motorway access to both the M4 and M5. Fourth, it had a great deal of adjacent undeveloped land available for distribution and port-related industries. However, they felt that as long as the National Dock Labour Scheme survived the potential of the port could not be realised. The abolition of the scheme in 1989 cleared the way for their bid, funded by loans from a number of banks and financial institutions.

The Council sought advice from accountants who confirmed the viability of First Corporate's business plan for the docks. There was little opposition from the port employees, and following a Council debate First Corporate acquired the land and assets of the port (excluding the old city-centre docks, now partly redeveloped for housing, leisure and tourism) on a 150-year lease on payment of a capital premium of £22.5 m. The city also acquired £11 m of shares in the company and agreed to transfer its responsibilities for navigation and harbour conservation to the company through a Harbour Revision Order (HRO), in return for which it would receive a further £2.5 m on completion of the transfer.

The deal was approved by Council with an overwhelming majority, although four left-wing councillors voted against or abstained, and the largest municipal port in the country passed into private hands in late 1991.

As elsewhere, however, privatisation proved controversial. Within weeks of the sale First Corporate announced a joint venture with National Power to build at Portbury one of the largest and most modern bulk handling terminals in Europe to handle, among other cargoes, the import of up to 7 million tonnes of coal per annum. This provoked strong opposition by the small group of left-wing councillors who had opposed the sale, and by the National Union of Mineworkers (NUM) who saw this as a threat to mining jobs in South Wales and the Midlands. The campaigners pursued their opposition through a stormy, three-week public inquiry into the granting of the HRO. They claimed that First Corporate had been a 'stalking horse' for National Power and that leading officers and councillors must have known what was in the offing when they

Figure 2.4. Bristol's coastal fringe. A view northwards in summer 1995 along the English shore of the Severn Estuary at its confluence with the Avon. South of their junction lies the latest (1977) dock enterprise serving Bristol, the Royal Portbury Dock, with its extensive car storage facilities (to the left) and bulk handling and storage units (above and right), along with the coal-conveyor serving the rail terminal in Avonmouth, north of the Avon. Avonmouth also contains the older outport developments of Bristol, and their associated industrial and warehousing zones. The old (1966) and new (1996) crossings of the Severn appear in the distance. (Courtesy of Bristol Port Company)

Figure 2.5.   The Royal Portbury Dock. Eventually opened in 1977 after a lengthy political battle, the Royal Portbury Dock is Bristol's latest outport development on the Severn Estuary coast. This summer 1995 view westwards across it to the Welsh shoreline in the distance shows clearly the extensive and still-expanding handling terminals for car imports and the new coal and other bulk goods terminal and storage bonds. The advantageous access to motorway (M5) traffic is also self-evident — there is a junction just off shot to the left. (Courtesy of Bristol Port Company)

recommended the sale of the port, even though no reference to large-scale coal imports was made at the time. From the other side, the directors of First Corporate were adamant that their bid did not depend on large-scale coal imports, and no deal had been struck with National Power prior to the sale, although 'we were of course aware that various bulk handling facilities were being discussed around the UK' (Letter to HRO Inquiry, 4 December 1992). They also pointed out that the terminal, though largely funded by National Power, would be used for other bulk cargoes as well.

The reason why National Power was prepared to pay so much for a terminal it might not use at anywhere near its capacity emerged in the reply given to a Parliamentary Select Committee Inquiry. When it was pointed out that if all the new terminals proposed in the UK were constructed they would have a total import capacity far in excess of likely imports, National Power claimed that up to a point it did not matter if these terminals were used or not. As long as the existence of the ports 'exercised . . . a discipline on the indigenous (coal) industry so that it is quoting competitive prices, then those ports have served their value to us quite as much as if physical coal had gone through them' (House of Commons, 1992, 31). An editorial in the trade paper *Lloyds List*, commenting on this attitude, referred to it as 'terminal madness' reflecting a 'mind-bendingly perverse logic' *Lloyds List*, 1993).

Initial evidence suggests that privatisation has brought considerable benefits to the port. Within the first year a number of major new projects were brought forward and total trade was reported to have increased by 19.3% between 1991 and 1993, a remarkable turnaround after a long period of decline. The coal terminal handles vessels of up to 100 000 tons, with coal going by rail to National Power's Didcot power station, although as a dual-use terminal this facility has handled more general bulk cargoes than coal. Elsewhere the port company's general strategy has been to make land available on favourable terms to encourage port users to invest, rather than investing heavily itself. Thus in 1993 United Molasses invested in the construction of what is claimed to be Europe's biggest warehouse for animal feed and fertiliser. There has also been a big increase in car imports, with a number of new terminal facilities handling Novas, Hondas, Fiats, Toyotas and Protons. The port has also attracted Mitsubishi and Colt imports from Newport and has now become Britain's largest car importing port. Finally, in 1993 the port also attracted a container line from Newport which runs regular services to Ireland and Spain.

Three factors seem to have been particularly important in attracting this increased trade. First, dock users have publicly stressed the importance of dealing with a dock management able to give almost instant decisions, compared to the laborious bureaucratic procedures of municipal control. Second, the workforce, though now down to 415, has been persuaded to adopt much more flexible labour practices, reducing turn-round times. Third, the availability of extensive areas of dockside land has proved attractive to users with large storage requirements.

In spite of this remarkable economic turnaround the direct impact of the port on local job prospects is likely to be limited by a number of factors. First, even if imports grow, with the increasing use of automated handling of bulk cargoes the number of dock employees is unlikely to increase much beyond the existing level. Second, although the traditional port-related industries such as the smelter and the fertiliser plant remain, their workforces have also been drastically cut and only about 10% of the port's traffic now services their needs (Bristol Port Company, pers. comm.). Third, the port management itself estimates that more than 90% of imports of timber, cars, coal and oil, and 65% of feedstuffs, passes directly out of the region for processing and consumption elsewhere.

As a result the port has appeared in local economic development and marketing strategies as a specialist adjunct to the local economy rather than a major driving force. These current strategies are partly the product of a remarkable transformation in local politics in the early 1990s. A new and more activist business élite has emerged and taken the lead in establishing a series of public–private partnerships, such as the Western Development Partnership

(WDP) which seeks to bring together representatives of all the major local authorities in the sub-region, together with leading businessmen and representatives of business organisations such as the Chamber of Commerce and the Confederation of British Industry. The WDP has commissioned various studies of the local economy and formulated an outline development strategy for the future (WDP, 1994). This future is seen to revolve primarily around sectors such as finance, aerospace, services and tourism. The port has been presented in strategy documents and publicity material as an additional specialist facility, giving Bristol a margin over its competitors. However, as we shall see, recent announcements suggest that the port itself may also become more of a driving force in the future.

A WIDENING AGENDA: PORT COMPETITION, COASTAL ZONE MANAGEMENT AND ESTUARY-WIDE PLANNING

The building of a coal terminal, the rapid expansion of dockside warehouses and the spread of car import compounds are indicative of the impacts of the revived port. However, these developments have intersected a swathe of other development pressures affecting the whole coastal strip from the port to the Severn Bridge. Similar kinds of pressures are also affecting coastal areas on the other side of the Severn. At the same time these development pressures are coming up against a revitalised environmental planning agenda for the later 1990s. As a result we have, on the one hand, the prospect of intensified competition between privatised ports, and on the other pressures towards greater cooperation between local authorities on both sides of the estuary to manage development pressures and their environmental consequences.

New development pressures in the Avonmouth/Severnside coastal zone

In 1995 the Avonmouth coastal zone, running from the port to the existing Severn Bridge, is being affected by a series of major development projects, some related to the port and some the result of other locational advantages (Figure 2.6). First, there are a series of major road schemes. The largest is the construction of the Second Severn Crossing which is due for completion in 1996 and will carry the M4 motorway to South Wales. A number of link roads are being constructed to new and existing industrial sites. As a result, large areas of land suitable for industrial development and warehousing will be opened up. ICI, which has a chemical plant in the area, owns much of this land and has held planning permission since 1957 for commercial development. A consortium of developers is seeking permission to develop a large site for industry, offices and distribution.

Second, outline planning permission was granted in 1995 for a 45-hectare rail terminal on land close to the port and industrial zone, permitting cargoes to be off-loaded straight onto trains, possibly for through transit to Europe. The port's existing container company (Bell Lines) has already announced its intention to

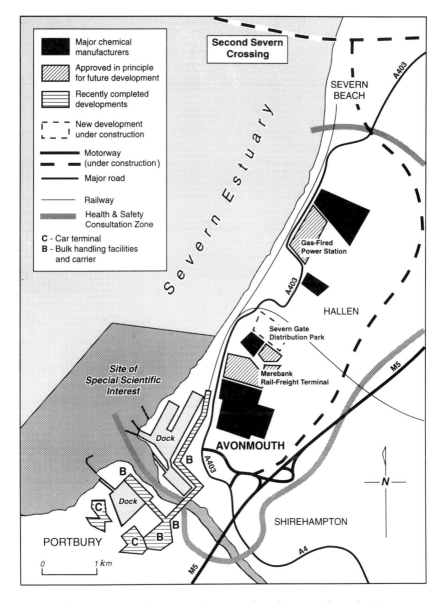

Figure 2.6. Development at Avonmouth and Severnside, mid-1990s

expand its operation and take over more dockside land; this rail facility can only make the port and its surrounding land even more attractive to potential port users and distributors. Third, the privatisation of the British electricity industry is having further impacts on the local area with the decision by Seabank Power to construct a new gas-fired power station at Avonmouth. Taken together with the new terminals and facilities linked to the port, all these schemes are generating a wide range of development pressures on the coastal strip in the mid-1990s, encouraging the local authorities in the area to cooperate in the formulation of a more comprehensive planning framework for the whole coastal zone.

Problems of coastal zone planning

Constructing an integrated plan has been made difficult in the recent past by a number of constraints. First, the national political context in the 1980s did not favour new planning initiatives. Although the UK's basic planning structure remained in place, the planning powers of local authorities were reduced, planning processes were speeded up, plans became less binding on developers and a greater reliance was placed on market processes (Thornley, 1991). Second, the coastal zone from the dock complex through the industrial zone to the site of the Second Severn Crossing has been the responsibility of four local authorities. Until recently these were the first-tier District authorities of Woodspring, Bristol and Northavon, and the second-tier County authority of Avon. These authorities have been under different political control (or, in the case of Avon, under no single party control), and have clashed on development priorities in the past. As for the future, central government accepted the 1994–95 local government review proposal that Avon should disappear from 1 April 1996, with the fate of the planning responsibilities it discharges in a state of flux at the time of writing.

Pressures towards more inter-authority planning also reflects a growing volume of environmental legislation. This is particularly important in the Avonmouth context as past industrial development has not only left a legacy of human health problems (particularly associated with heavy metal pollution from the smelter), but also one of extensive land and water contamination. Yet adjacent coastal zones contain a number of Sites of Special Scientific Interest and the whole upper Severn Estuary is an environmentally valuable and ecologically sensitive area.

However, the environmental legislation affecting coastal zone planning is bewildering in its complexity. The latest government planning guidance note on 'Coastal Planning' (DOE, 1992) lists eight central government departments and 13 other bodies with some involvement or responsibilities in the policy area. In the same year the House of Commons Public Accounts Committee identified 160 bodies concerned with coastal zone defences, commenting on the 'complexity and potential confusion' which led to a 'piecemeal approach and inhibited a wider strategic view'. In terms of legislation, over 80 Acts of Parliament seek to regulate coastal zone activities, to which have now been added three important EU directives (MAFF, 1993). Small wonder that Carter (1988) considered that, in comparison with practice overseas, British coastal management is 'cumbersome, muddled and unlikely to be very effective'.

In this context the Severnside local authorities have nevertheless begun to make some progress. Bristol's draft District Plan (Bristol City Council, 1992) accorded the Avonmouth area a chapter to itself, with a strong environmental flavour. This plan in turn was linked to a longer-term (20–30 year) strategy plan for the whole Avonmouth/Severnside area recently drawn up by a joint planning team from three of the local authorities. This recognises that the area provides the best opportunity for the encouragement of manufacturing industry in the former county, but the kind of industrial development that will cause further environmental deterioration is to be resisted, green corridors are to be preserved, and the coastal zone and its ecology protected.

Estuary-wide planning: the emerging European context

Developments at Avonmouth have to be seen in the context of coastal zone development pressures in the Severn Estuary as a whole. The estuary has long been an arena for grand projects. The 1971 Severnside Report noted proposals for new motorways, a second motorway bridge, an international airport on reclaimed mudflats, Bristol's Portbury project, and a series of possible Maritime Industrial Development Areas. Some of these schemes have since been completed, others have fallen by the wayside. Overshadowing them all, however, has been the proposed Severn Barrage which has been the focus of a whole series of reports and studies since 1910. The specific barrage favoured in the 1980s was intended to generate power from the huge tidal range in the estuary (the second highest in the world). It would be 10 miles (16 km) long, cost £9.4 billion to build (at 1988 prices) and produce 7% of Britain's electricity needs. It would also have immense economic and ecological impacts on the estuary, some of which were explored in the 1980s through a series of government-sponsored technical studies.

The sheer scale and complexity of these impacts brought local authorities on both sides of the estuary together in the early 1980s in a joint organisation called SCOSLA (Standing Conference of Severnside Local Authorities). One of its concerns was the impact on the three major ports which would find themselves upstream of the barrage, and they campaigned for locks to be provided in the barrage capable of handling the largest ships currently using those ports. Further technical studies suggested that delays in using the locks would be largely outweighed by the extra time in which ships could freely enter the upstream ports on the higher tides.

In the event, the local authorities proved somewhat divided on the merits of the scheme, but this is less of a problem now that any government commitment to the project is unlikely in the near future. Nevertheless, the experience of working together has been sufficiently valuable for the local authorities involved to consider retaining SCOSLA and expanding its remit to include broader issues of estuary-wide planning. This recognition of the need for more regional cooperation has been prompted by the rush of other development proposals, including the Second Severn Crossing, the Cardiff Bay barrage, a new international airport on the mudflats between Newport and Cardiff, and new motorway links on the Welsh side of the Second Severn Crossing.

This focus on the estuary as a whole has now been reinforced by a Europe-wide initiative. The willingness of local authorities to become involved in European projects in alliance with European authorities and agencies is a new phenomenon, at least on the Bristol side of the estuary. One recent example has been the involvement of the Western Development Partnership in drawing up a successful bid for defence-industry conversion funds under the EU's KONVER programme to combat the effects of closing defence plants and military bases. The involvement of the local authorities in the area in the ESTURIALES initiative is another sign of this changing attitude towards Europe.

The ESTURIALES grouping was initiated by the Association Communautaire de L'Estuaire de la Loire, a grouping of French local authorities in the lower Loire valley, and a subsequent conference in 1991 brought together the local authorities representing the Loire, Clyde, Severn, Tagus and Wear estuaries. A series of major studies of each estuary was launched using a common format, with a view to bidding for funds under the European Union's LIFE programme (Esturiales, 1993). The general aim of ESTURIALES is to bring about the 'sustainable development' of each estuary in a way that does not compromise the ability of future generations to meet their own needs. Sustainable development requires taking a 'holistic' view of each estuary, integrating economic and ecological appraisals in such a way as to avoid the kind of developments that would bring about irreversible damage. This in turn requires the formation of estuary-wide planning and management systems.

The Severn Estuary study follows these lines, bringing together a wide range of existing information on administrative structures, proposed developments, environmental and pollution problems, and coastal processes. Not surprisingly perhaps, the study concludes that the Severn is subject to fragmented and *ad hoc* decision-making by a multiplicity of public and private bodies. It proposes the establishment of a Forum of estuary users, including port interests, and an Estuary Management Group to develop an integrated Estuary Zone Management Plan (EZMP). The study has formed the basis of a bid for European funding.

These estuary-wide plans represent an interesting and potentially very important step forward, considerably raising the profile of environmental constraints on future industrial and port development. However, it remains to be seen whether such an initiative can overcome traditional conflicts of interest which in the past, as we have seen, have pitted South Wales against Bristol. As long as South Wales has the advantage of Development Area status and a Welsh Development Agency dedicated to securing advantages for South Wales industries and ports, it will be difficult to refocus planning on the integrated and balanced development of the estuary as a whole. In a sense we have an interesting tension developing here between the 1980s trend towards a more competitive and market-orientated environment for industrial location and port development, and what may be the emerging trend of the 1990s towards building inter-regional alliances and cooperation in order to secure European funding. It will be fascinating to see how the ports respond to this wider framework.

CONCLUSIONS

Our general theme has been the changing relationship between port and city in the context of regional and national change. It is evident that privatisation has already had important effects on the prosperity and prospects for different ports in the Severn Estuary. It has also resulted in a re-casting of relationships between port and city, with ports achieving greater autonomy as private companies. However, the sheer scale of industrial and port-related developments in some parts of the region is forcing local authorities to find new ways of cooperating with each other to develop integrated environmental strategies up to the estuary-wide scale. Future port-related development will have to conform to this emerging environmental agenda.

REFERENCES

ABP (1990), *Annual report and accounts* (London: Associated British Ports plc).

Asteris, M. (1988), 'Britain's seaports: competition and trans-shipment', *National Westminster Bank Review*, February, 30–48.

Avon County Council (1993), *Avonmouth/Severnside draft strategy* (Bristol: Avon County Council, Bristol City Council, Northavon District Council).

Bassett, K. (1986), 'Economic restructuring, spatial coalitions, and local economic development strategies', *Political Geography Quarterly, Supplement* 5(4), 163–78.

Bassett, K. (1993), 'British port privatisation and its impact on the port of Bristol', *Journal of Transport Geography* 1(4), 255–67.

Bassett, K. and Hoare, A. (1984), 'Bristol and the saga of Royal Portbury: a case study in local politics and municipal enterprise', *Political Geography Quarterly* 3(3), 223–50.

Bird, J. H. (1963), *The major seaports of the United Kingdom* (London: Hutchinson).

Boddy, M., Lovering, J. and Bassett, K. (1986), *Sunbelt city?* (Oxford: Clarendon Press).

Bristol City Council (1992), *Draft Bristol local plan: written statement* (Bristol: Directorate of Planning).

Carter, R. (1988), *Coastal environments* (London: Academic Press).

Committee of Public Accounts (1994), *Department of Transport: the first sales of trust ports*, 31st Report, House of Commons Service 1993–94 (London: HMSO).

DOE (1971), *Severnside: a feasibility study* (London: Department of the Environment and Welsh Office, HMSO).

DOE (1992), *Planning Policy Guidance 20: Coastal Planning* (London: Department of the Environment and Welsh Office, HMSO).

Esturiales (1993), *Environmental study* (Swansea: Association Communautaire de l'Estuaire de la Loire, Swansea Institute of Higher Education).

Financial Times (1992), 'Ports in disarray', *Financial Times*, 14 January, 14.

House of Commons, Trade and Industry Committee (1992), *British energy policy and the market for coal: minutes of evidence*, 10 November (London: HMSO).

HRO (Harbour Revision Order) (1992–93), *Enquiry transcript and evidence* (Bristol: Bristol City Council).

Lloyds List (1993) 'Terminal madness', *Lloyds List*, 3 February, 5.

MAFF (1993), *Coastal defence and the environment* (London: Ministry of Agriculture, Fisheries and Food).

Morgan, K. (1993), *Bristol and the Atlantic trade in the eighteenth century* (Cambridge: University Press).

Sacks, D. (1991), *The widening gate: Bristol and the Atlantic economy, 1450–1700*

(Berkeley: University of California Press).

Thornley, A. (1991), *Urban planning under Thatcherism* (London: Routledge).

Turnbull, P. (1991), 'Labour market de-regulation and economic performance: the case of Britain's docks', *Work, Employment and Society* 5(1), 17–35.

Turnbull, P. and Weston, S. (1992), 'Employment regulation, state intervention and the economic performance of European ports', *Cambridge Journal of Economics* 16, 385–404.

Turnbull, P. and Weston, S. (1993), 'Co-operation or control? Capital restructuring and labour relations on the docks', *British Journal of Industrial Relations* 31(1), 115–34.

WDP (1994), *Avon economic development strategy: consultation draft* (Avon: Western Development Partnership).

3 Managing the Cityport/Coastal Zone Interface: A Mersey Estuary Case Study*

DAVID MASSEY

Department of Civic Design, University of Liverpool, UK

Since the 1970s there has been a growing concern with the changing relationship between cityports and the wider coastal areas/zones of which they form part. This concern has both analytical and policy aspects. Its essential features are the wider perspectives of the coastal area it gives and the emergence of new management goals reaching for oversight of the traditional primacy of economic growth and its particular expression from the 1970s in the redevelopment of the waterfront (Hoyle *et al.*, 1988). Vallega (1992, 11) has more recently described how

> ... the littoral region tends to be perceived in a different way from the past: now it is regarded more and more as a coastal management area, i.e. as an area consisting of land and a marine zone the resources of which are to be exploited as a whole and the environmental protection of which is to be pursued as a main objective.

In the policy field new institutional arrangements and management systems (Woodcock and Keen, 1994) have been brought into being in, for instance, North America, Europe and Australasia to address these wider relationships and new goals and objectives in a more integrated way. Although the specific organisational aspects differ from place to place (OECD, 1993), the coastal zone management objectives pursued in such initiatives all focus to greater or lesser degree on the sustainable use of coastal resources as their guiding principle.

In Britain it is possible to discern several related clusters of activity in terms of coastal zone planning and management (King and Bridge, 1994). Among these are attempts to define national concerns and guidance (e.g. Department of the Environment, 1992a, b; House of Commons Environment Committee, 1992). Another cluster consists of regional and local-level efforts to devise strategic policies for particular administratively-defined sections of coast (e.g. East Sussex

---

* This chapter is written in a personal capacity and the views expressed in it should not necessarily be read as expressing those of the Mersey Estuary Project Group.

County Planning Department, 1992; SERPLAN, 1993). A third cluster comprises a series of estuary-based initiatives including the major programme promoted by English Nature (1992, 1993), joint exercises covering the Severn and the Dee estuaries, and, Scottish Natural Heritage's 'Focus on Forths' initiative. The estuary-based initiatives form a precise focus for examining the dynamic relationships between cityports and their immediate coastal areas in the light of new management goals and objectives, with the earliest such project for a major urban–industrial estuary being established for the Mersey Estuary in North West England in 1992.

This chapter first describes the Mersey Estuary Zone (Figure 3.1), its seaport system and its cityport extending beyond the conurbation of Merseyside to take in a wider sub-regional urban–industrial structure. It then turns to explore the development of changes over the last 30 years in the economy of the seaport system, in water quality concerns and in nature conservation values. In the 1980s, together with the establishment of waterfront and urban regeneration programmes, each of these policy areas received increasing if somewhat uncoordinated attention. In these circumstances, the perceived threat of major infrastructure proposals in the late 1980s to early 1990s prompted the commissioning of the Mersey Estuary Management Plan exercise in 1992. The chapter then examines the organisational arrangements made for the Management Plan, its work programme and its proposals for the future. It concludes with a brief review of some leading issues for management plan implementation processes.

THE MERSEY ESTUARY ZONE AND CITYPORT COMPLEX

The Mersey Estuary provides the outlet to the Irish Sea for the Mersey Basin group of rivers – which also includes the Irwell, the Tame, the Goyt, the Weaver, the Gowy and the Alt among other watercourses. It drains a heavily industrialised catchment of some $5000 \, km^2$ with a population of 4.5 million including the metropolitan areas of Merseyside and Greater Manchester.

The estuary can be divided into four sections (Porter, 1973). The Upper Estuary from Howley Weir (the normal tidal limit) in Warrington to the Runcorn–Widnes Gap is a narrow meandering channel set in a continuation of the river's floodplain which extends in a non-tidal section to the east of Warrington. Beyond Runcorn Gap the estuary opens out to the south and west into the broad (4–5 km wide) and shallow basin of the Inner Estuary with extensive sand and mudflats at low water and extensive saltmarshes on its southern flanks. From Eastham–Garston the estuary begins to narrow, its banks densely built-up with docks, industry, housing and reclaimed open space. The Narrows section continues to the mouth of the estuary between New Brighton and Crosby where it opens along the sand-dune coastlines of Liverpool Bay to

---

Figure 3.1.   (*opposite*) The Mersey Estuary Management Plan study area

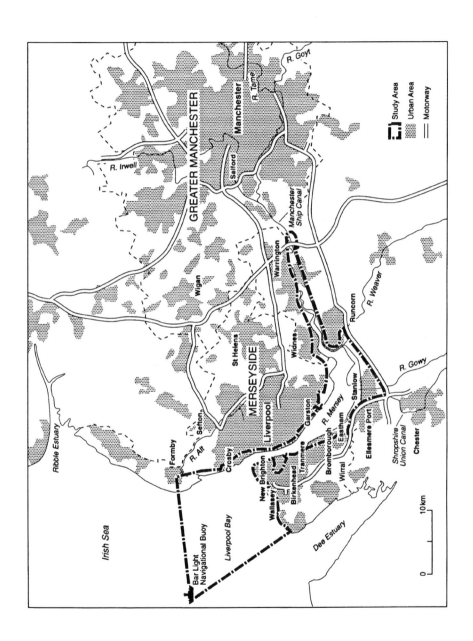

both the north (towards the Ribble Estuary) and to the west (towards the River Dee Estuary) (Rice and Putwain, 1987). The river then makes its way to the Irish Sea through the training banks of the Crosby and Queen's channels. From this position the Mersey Estuary looks outward to the North Atlantic and beyond providing the locus for a seaport system with extensive landward connections, well placed to service both short sea routes to Ireland and mid-sea/deep-sea European and inter-oceanic routes (Hyde, 1971).

The Mersey estuary seaport system

The development of the seaport system has had to contend with two natural factors which give it a distinctive quality. The tidal range of up to 10 metres is extensive and, together with the effects of wind and of currents of up to 6 knots through the Narrows, made for the early (1709–15) establishment of a wet dock system within existing tidal creeks (as with the original Old Dock, the Birkenhead–Wallasey Docks and at Bromborough Pool) or more commonly built out into the river itself, as with the south Liverpool (Ritchie-Noakes, 1984) and the north Liverpool–Bootle–Crosby docks. The second factor has been the need to provide dredging and training banks for a regulated deep-water channel from the Narrows to the Irish Sea (Bird, 1963), a need which became acute with the growth in the size of transatlantic passenger liners in the late 19th century.

Although the establishment of the Mersey Docks and Harbour Board (Mountfield, 1965) in 1857 to control the Birkenhead and Liverpool docks provided the estuary with its leading seaport operator, other smaller operators maintained services in the Inner and even in the Upper Estuary. These included Garston, now owned by Associated British Ports (ABP), Bromborough Dock, opened by Unilevers in connection with their Port Sunlight works in 1931 (Bird, 1963) and closed in 1986 (ACRM, 1986), Weston Point and Runcorn. However, just as a seaward channel was important for continuing commercial navigation access to the Narrows and beyond, so too have landward links (by road, rail and canal) been significant in maintaining access to the estuary's hinterlands.

The most important canal development in terms of the Mersey Estuary seaport system in the last century or so has been the opening of the Manchester Ship Canal in 1894 (Bird, 1963). This project had the effect of extending the estuary's existing seaport into a more complex system, as it developed into an extended port and not just a connection to its inland terminal in Manchester–Salford. In particular, docks and wharves were built in connection with the industrial areas along the line of the Lower Canal around Ellesmere Port (terminal of the Shropshire Union Canal) and the Stanlow petrochemical complex (Aspinall and Hudson, 1982) and ICI's heavy chemicals plants in Runcorn (Warren, 1980). Shell UK's oil refinery at Stanlow has played a particularly significant seaport development role with oil handling facilities being constructed at Stanlow and Eastham (Queen Elizabeth II Dock) and an oil terminal (with a pipeline link to Stanlow) built in the Narrows at Tranmere. Use of the terminal for the bulk import of crude oil was superseded for a time by a 100-km pipeline to an offshore mooring point at Anglesey to handle Very Large Crude Carriers (VLCCs), but it was upgraded and re-opened in 1988, importing up to 12–13 million tonnes a year. In August 1989, 150 tonnes of crude oil were discharged into the estuary when the Tranmere–Stanlow pipeline fractured, giving a focus for serious ecological and environmental protection concerns (EAU, 1991).

The Mersey Estuary cityport

The spatial structure of the Mersey Estuary cityport reflects its relatively complex seaport system. Its hub lies in Liverpool city centre with its office-quarter core of late Georgian–Victorian office buildings adjacent to Pier Head and its retail core around Church Street. The city centre serves not only Liverpool but also a wider Merseyside metropolitan area on both banks of the estuary in Sefton and Wirral and an Estuary Zone sub-region including Widnes and Runcorn, Warrington, Stanlow and Ellesmere Port related to the Inner and Upper sections of the estuary and the Manchester Ship Canal. In addition to the industrial and urban settlements of the extended sub-region, the Estuary Zone contains two power stations owned by Powergen plc at Fiddler's Ferry, between Widnes and Warrington, and at Ince near Stanlow. In mid-1995 a third major power station for the Estuary Zone was proposed by ICI, in association with a United States' consortium, for a site at Runcorn to serve its local energy-intensive, chlor-alkali plants.

CHANGING PERSPECTIVES

The Mersey Estuary had provided the location for the development of an internationally ranking seaport system and an industrialised and urbanised sub-region with relatively little concern for its natural resources. But by the 1960s a turning point had been reached in many respects in the relationships between the seaport, the urban–industrial sub-region and the water body of the estuary. New directions for change emerged in the 1970s and 1980s with a dramatic change in the ecological and nature conservation value of the Inner Estuary and the growth of environmental protection considerations in public policy which impinged on those relationships and brought a new dimension to them.

Turning points

Gilman and Burn (1982, 17) have described how 'the Port of Liverpool was a giant of the age of conventional general cargo shipping which lasted until the middle of the 1960s'. The Mersey Docks and Harbour Board's response to the changes in transport technology of the introduction of container shipping was to convert its mid-1960s plans for a large conventional expansion at Seaforth into a deep-sea container terminal with specialist bulk facilities, which was formally opened in 1973. The £50 m project put a considerable strain on the Board's finances and it was re-constituted as the Mersey Docks and Harbour Company (MDHC) (Lynch, 1994). At the same time the 95 hectares of land and water space of Liverpool South Docks were closed to traffic, with dock basins, such as the Albert Dock, left to silt up. A decade of frustration ensued, in which ambitious plans and development proposals were suggested but nothing achieved and the physical structure of the system continued to deteriorate.

In terms of the estuary's water quality, the middle and late 1960s also marked a turning point of sorts, in that the pollution load carried is estimated to have reached its maximum. Two problems were identified: first, offensive smells in the Upper Estuary in hot summer months, the result of an almost complete lack of dissolved oxygen (Buckley, 1980, 322–3): and, second, 'objectionable gross solids of sewage origin' in the Upper Estuary with evidence on the shoreline and beaches of the lower and outer parts of the estuary of sewage waste agglomerated with grease, fat and oils (Alexander, 1982). Consultant's reports and scientific enquiries including the construction of a mathematical model of the estuary followed public protests and national publicity. The inauguration of a new regional water authority for the North West region in 1974, including the whole of the Mersey Basin, provided the essential institutional structure to organise a £150 m programme to achieve the minimum objectives of smells and solids removals. However, given its other, wider commitments, the Authority could only see its way to funding an initial 'package' of £50–60 m over the next 15 years (NWWA, 1980).

In the late 1960s the Mersey Estuary (more particularly the Inner Estuary) was considered to be of very limited ecological importance (Cheshire County Council, 1989), but by the later 1980s it had risen to fifth in the Nature Conservancy Council's national rankings, principally because of its large overwintering population of wildfowl. Among the factors suggested for this dramatic change from the early 1970s are the displacement of large flocks from wintering grounds elsewhere in Europe, especially from the north and west of the Netherlands; the availability of an undisturbed and underused habitat protected along its southern margins by the Ship Canal; the beginnings of some improvement in water quality; an increase in the area of saltmarsh providing roosting and feeding areas; and, the availability of abundant food sources probably helped by the river's high organic load (Cheshire County Council, 1989). The Inner Estuary's changed nature conservation status led to its recognition as a Site of Special Scientific Interest (SSSI) along with similar designations for the Alt Estuary and the North Wirral Foreshore in the Outer Estuary and the Ship Canal Company's dredging deposit grounds at Woolston in the Upper Estuary.

New directions

The 1980s saw not only changes in nature conservation values in estuary management policies, but a considerable improvement in the trading fortunes of the MDHC from its rather bleak outlook in 1982 (Gilman and Burn, 1982), marked not just by the abolition of the National Dock Labour Scheme in 1989, but by rationalisation projects and developments designed to maintain and attract new traffic (MDHC, 1989). Among these were the establishment of a successful Freeport Zone in 1984 and the initiation of several development projects on the Dock Estate including a business park on the site of former lairages at Birkenhead Woodside and the conversion of former warehouses with a dock frontage into residential apartments. Although at one time it had looked as if the MDHC would lease the Liverpool South Docks to Merseyside County Council for a metropolitan-scale regeneration programme, the opportunity was removed by the establishment of the first of the government's urban development corporations in 1981.

The Merseyside Development Corporation (Adcock, 1984) quickly began to implement an initial strategy (MDC, 1981) of reclaiming the silted water space in the docks and refurbishing the better buildings for re-use. The 'flagship' scheme in this process was the Grade 1 listed Albert Dock group of buildings, containing retail outlets and small office suites, the Merseyside Maritime Museum, the Tate Gallery North, apartments and a regional television news headquarters. It was complemented by the 1984 International Garden Festival scheme, which transformed the riverside municipal waste disposal area into a new complex of visitor attractions, open space and housing development sites. In 1988 the Corporation's designated area was almost trebled in size with new development areas in Liverpool, Birkenhead and New Brighton (MDC, 1990).

The Mersey Docks and Harbour Company had also begun to look for new users and new uses for its North Liverpool Docks as it rationalised its operations into the Birkenhead–Wallasey Docks and the Bootle–Crosby Docks, together from 1988 with the renovated Tranmere Oil Terminal. For the 14.6-hectare Princes Dock site immediately adjacent to Pier Head, it sought private sector partners for a commercial office–residential–hotel scheme, establishing a joint venture with P & O Developments, and proposed to make a start on the 116 125 m² scheme with a modest 4465 m² office development in late 1995. Further north, however, at Sandon Dock the Company had more success in finding a new user as the North West Water Authority sought an appropriate site for a new sewage treatment works linked to its 1980 initial 'package' proposal for a Crosby to Speke intercepting sewer, which would also serve as a reception point for a sludge pipeline from its Davyhulme works in Manchester, and a sea terminal for sludge disposal (Dixon, 1985). The new project also brought changes for the Ship Canal Company, since not only had it lost traffic in the upper reaches of the canal, but the sludge pipeline would bring about the further loss of its largest remaining user (MSCSC, 1985). In the event the Company's proposal to close the upper reaches to navigation was averted, but it emphasised the changed nature of the canal from its origins and the new focus on its lower reaches from Ellesmere Port to Runcorn.

The Mersey Basin Campaign

Although its commitment to implement the intercepting sewer scheme finally earned North West Water the planning permission it needed from Liverpool City Council for the Sandon Dock project to go ahead, the Authority was very much aware of how capital shortages were holding it back from the full programme needed. Arising from his commission as Minister for Merseyside in the wake of the civil disturbances in Toxteth in 1981, the Secretary of State for the Environment, Michael Heseltine, provided the national political backing first to provide substantially increased funding for the full £170 m programme over the same 15-year period (Dixon, 1985) and, second, to launch '... a campaign to bring about a dramatic and lasting improvement in the state of the Mersey and its banks' (Heseltine, 1982). From this initial proposal the Mersey Basin Campaign emerged in 1985 as a 25-year, £2 billion water quality improvement programme, which incorporated the expanded Mersey Estuary Pollution Alleviation Scheme and set out the wider complementary objective of clearing away the dereliction on the landward side through 'imaginative schemes for riverside development – recreational, residential, commercial and industrial' (Tavare, 1986, 2).

TOWARDS THE MANAGEMENT PLAN INTERFACE

Proposals for new infrastructure projects

Although Merseyside County Council was not to lead the regeneration of the Liverpool South Docks, this proposal was not the only strategic opportunity it had under consideration. In late 1980 a number of ideas for a series of wide-ranging projects emerged at a conference convened by the Council to explore ways of bringing back prosperity to the estuary, its seaport and cityport (Chartres, 1980). The chief executive of the MDHC, James Fitzpatrick, suggested that a tidal barrage might be constructed at the mouth of the river between New Brighton and Crosby, which could be used to generate electricity and make possible the construction of new ship terminals which would not need to be entered by locks. Others present saw the vast ponded area created with its diminished tidal range as a recreation opportunity, being filled with marinas at prices highly competitive with those in the south of England. Mr Fitzpatrick also suggested that a new international airport could be built in the Outer Estuary on a reclaimed Great Burbo Bank off the north shore of the Wirral coastline. Further discussion about the airport idea did not proceed, although ironically a decade later British Aerospace, the majority shareholder in Liverpool's Speke Airport located on an Inner Estuary site between Garston and the greenbelt land of the Oglet and Hale peninsulas, came forward with its own proposals for expansion including the concept of a new 40-million passenger capacity international hub airport involving very substantial land reclamation from the river in phases beyond 2005 (Brown, 1990).

On the other hand, the barrage idea was taken forward more immediately with enthusiasm tempered with some scepticism, first by the County's Enterprise Forum in 1981, followed by a pre-feasibility study commissioned by the Council whose 1983 report was itself followed by further economic, hydraulics and engineering studies. In 1986, the year of the abolition of the metropolitan county councils, the proposal was put on a separate footing with the founding of the Mersey Barrage Company; and, with support from the Department of Energy's tidal resource programme, work began on a detailed technical feasibility study into where to construct the barrage, how it would affect tides, sedimentation, ecology and ship movements, and what its social and economic impacts would be. Two interests increasingly saw the proposal as more of a threat than an opportunity: first, and somewhat ironically in view of the source of the initial proposal, the port operators, shipping and commercial bodies, such as Shell UK, with a concern for commercial navigation, and, second, national and local ecology and nature conservation bodies.

The barrage was seen by the first group as producing potentially harmful and unforeseeable changes in the estuary's navigation channels and, through the need to use locks to enter and exit the ponded area behind the barrage, as increasing the costs and time of commercial shipping. The ecology group saw the barrage as threatening the new-found national and international status of the Inner Estuary through the diminution of the tidal range in the ponded area with its consequences for the feeding resource and also through its effects on the channel and land–waterspace structure of the ponded area. The Barrage Company's approach to considering the ecological impact of its proposals included liaising with the Mersey Barrage Ecology Consultative Group, which in the course of its work began to put together its concerns about the more extended range of changes which might have potential impacts on the estuary above and beyond those related to the barrage. These included the Mersey Basin Campaign's pollution alleviation scheme, with its potential for restoring commercial fishing and opening up new or restored prospects for recreation along the shoreline; the further potential for tourism and recreation demonstrated by the Merseyside Development Corporation's waterfront regeneration programme; and ideas for an impounded Water Park leisure lagoon upstream of Runcorn Bridge proposed by Halton Borough Council (Handley, 1992).

Initial ideas for a Mersey Estuary management plan

Given the potential impact of these proposals together with the proposal for the barrage and the prospect of increasing competition between uses and users of the water frontage and the water surface, the Consultative Group put forward in 1989 the idea of a Management Plan for the Mersey Estuary which would 'provide a framework for the management of the Mersey Estuary within which existing interests can be safeguarded whilst realising the full potential of the Estuary as a natural resource' (Handley, 1992). With this proposal came the decisive conceptual step in recognising the interconnected set of relationships which had been developing in the specific context of the Mersey Estuary in the 1980s and the definition of 'a new planning subject' (Vallega, 1992, 115), designed to address environmental as well as economic development concerns. If the seaport/waterfront is the functional and physical interface between the cityport and its water body, then the coastal/estuarial management plan can be seen as the conceptual and policy interface.

Although there were no specific precedents to draw on, John Handley, Chairman of the Ecology Consultative Group, suggested that the Nature Conservancy Council's site management plans approach for nature conservation and the National Park Plans produced by National Park authorities in England and Wales might provide initial leads. His paper set out the need to define an Estuary Zone 'well into Liverpool Bay' and extending beyond the normal tidal limits at Howley Weir through Warrington as far as the administrative boundary with Greater Manchester, to involve a wide range of local authorities, regulatory and managing agencies, owners, occupiers and users of the estuary and special interest groups. Among the representative bodies he thought might help in facilitating the extensive and lengthy consultation needed was the Mersey Basin Campaign's Estuary Project Group.

Taking the practical step

The Estuary Project Group was one of a number of similar territorially based bodies established by the Campaign (MBC, 1986) to bring together representatives of the North West Water Authority with local authorities, public agencies and private and voluntary interests for liaison and new local initiative purposes. The Project Group's response to John Handley's paper went beyond simply facilitating consultation and took more positive form in the creation of six topic groups coordinated by Liverpool City Council (who chaired the Project Group) in late 1990 to identify the main issues and conflicts which would need to be addressed. The topic groups covered the subjects of: fisheries, land drainage and coastal protection; pollution and hazards; conservation; navigation; development and transport; and recreation. By the time the topic groups had completed their work, a further development proposal had been identified in the Department of Transport's decision to commission a study of cross-Mersey travel demand and the role that might be played by a third river crossing.

Having identified the leading issues and conflicts, the Project Group went a stage further and in 1991–92 promoted the commissioning of a three-year exercise to prepare a management plan from a University of Liverpool team who, although locally based, would be able to provide an informed yet neutral perspective (Batey and Kidd, 1993, 29).

THE MERSEY ESTUARY MANAGEMENT PLAN

Commissioning the Plan

While the overall membership of the Mersey Basin Campaign's Estuary Project Group provided a broadly representative body responsible for oversight of the preparation of the Management Plan, financial support for the exercise came substantially from the Department of the Environment (Mersey Basin Campaign Unit) together with contributions from English Nature, the National Rivers Authority (which had taken over the North West Water Authority's regulatory functions on privatisation in 1989) and the Estuary Zone's local authorities (Batey and Kidd, 1993). More direct and regular guidance and support to the university team was provided by a Technical Steering Group, chaired by a member of Cheshire County Council's Planning Department, with a membership based on local knowledge and specific expertise. The University of Liverpool team drew on the involvement of four members of the academic staff of the Department of Civic Design and one member of the Department of Politics, together with part-time research assistance from some 15 graduate students over the three-year period of the commission from 1992 to 1995.

The study brief took the 1989 Handley report's general objective as its starting point and specified this more closely so that the Management Plan would:

- focus attention on the Estuary as one of the Mersey region's most important environmental assets and convey a positive image of the area as a unique conurbation with an enormous water resource (with recreational and tourist potential) at its core;
- provide the basis for an agreed and coordinated programme of environmental action and creative conservation to be implemented by the commissioning partners and others;
- set out proposals for the management of river-based recreation and for the protection of ecological assets;
- establish part of the technical basis to enable the local authorities and others to respond to major development initiatives in the Estuary;
- enable the commissioning partners to speak with an informed and authoritative voice on matters affecting the Estuary.

The existing urban–industrial nature of the Estuary Zone was seen as being implicit in the first three of these objectives, with the emphasis being placed on the new functions of tourism and recreation and environmental protection and nature conservation as leading issues where overviews and coordinated approaches were particularly needed. The task for the team was to bring these aspects together with existing and proposed policies for more well-established functions (e.g. water quality improvement, navigation and port operation) and the development plans and policies of the waterfront local authorities, thus meeting the overall requirements of the general objective and the more technical and specific requirements of the fourth and fifth more detailed objectives.

Defining the Estuary Zone

Among the first tasks to be undertaken was the definition of the study area (the 'Estuary Zone'). The approach again took its lead from the Handley report with the Bar Light navigational buoy in Liverpool Bay marking the westward and seaward limit and the Ship Canal's Woolston dredging deposit grounds and the M6 motorway marking the eastward limit. In the Outer Estuary the Wirral coastline was treated as a joint oversight responsibility with a parallel exercise being initiated for the Dee Estuary (Dee Estuary Forum, 1994) under the auspices of English Nature's programme, while in Sefton the northerly limit at Formby Point linked with the well-established Sefton Coast Management Scheme, which provided a link through to the emerging Ribble Estuary Management Study. The general landward limits of the study area were defined partly with reference to the 10 m contour and partly with reference to well-established transport and built environment features. The result was very much an *ad hoc* boundary, and it was recognised that some functions and aspects would need to be considered beyond the immediate confines of the defined boundary and that the scale of the estuary required some treatment at an area basis within the zone as a whole.

Method of working

The work programme for the preparation of the Management Plan was structured around four annual conferences, designed to bring issues and proposals before a generally representative audience: the first conference (1992) was used to launch the Plan preparation process; the second conference (1993) was used to report on the study team's initial technical work; the third conference (1994) provided an opportunity for the team to present their initial proposals for comments and discussion; and the fourth conference (1995) was used for a presentation of the outcomes of the consultation process on the initial draft of the Plan which had been issued in October 1994, and for discussion on the organisational arrangements to be made for the Plan's implementation after its formal publication in late 1995. The study team was not asked to bring forward recommendations on the latter item, but provided an advisory input through a review of national and international experience (Woodcock and Keen, 1994).

The annual conferences were complemented by a series of specialist surveys of statutory authorities, voluntary, community and private sector interests, and a series of workshops in Warrington, Chester and Bootle taking a more local area-based approach to issues, options and preferred policy directions. Specialist technical studies were made of land ownership and tenure, navigation, tidal regime and level of use, water quality and nature conservation, formal, informal and water-based recreation, tourism, emergency planning, the implications of the EC waste water directive, and fisheries and fishing. These studies were supplemented by a separate project commissioned by English Nature examining sedimentation patterns in the estuary and by a report on environmental quality by the National Rivers Authority (1995).

Identifying issues

The study team's analysis of the issues and conflicts emerging from the technical and specialist studies and various consultations identified four strategic and highly inter-related areas of concern in terms of the subjects of management processes. The first identified strategic issue was the protection, management and enhancement of the natural and man-made resources of the Estuary Zone, a definition which extended the original concern with natural resources to take on a more holistic approach. This issue was more closely specified in terms of estuary dynamics, water quality and pollution control, biodiversity, and land use and development. It reflected expressed needs for better understanding of the complex processes and interactions which continue to shape the estuary environment. Reflecting the longstanding characteristics of the economy of the Merseyside metropolitan county, a characteristic sharply emphasised by its 1994 designation as the European Union's first older industrial area for Objective 1 regional assistance, the second strategic issue was economic development. Closer specification of this issue identified commercial navigation and port development, urban regeneration and tourism development as key aspects.

Recreation provided the third strategic issue for consideration, with its specialist aspects of informal and shore-based recreation, and water-based sport and recreation being seen as closely connected with the tourism issue in the economic development issue area. The fourth issue identified was the need to promote an appropriate integrated monitoring of estuary conditions so as to assist in tracking the implementation of policies and to provide information alerting managers to unforeseen changes, and to promote a deeper and wider understanding of the estuary and efforts to provide for its management through a wider and more open consideration of issues, information and policy options.

Developing the policy framework

In addressing the development of management policies to meet the identified strategic issues and their more closely specified policy areas, the study team worked on the basis that it needed to develop policies which would support and draw on existing management processes as well as suggest new features and proposals for change. In that sense, the Management Plan came to be developed as an advisory strategic guidance document designed to meet the requirements of an evolving set of inter-organisational management decisions and processes, rather than a site-based, project-specific action programme or a statement reinforcing rigid area protection designations. The team also wished to leave scope for good connections to be made with similar exercises along the North Wales–North West England region of the Irish Sea coast; with the Ministry of

Agriculture, Fisheries and Food's guidance on flood and coastal defence which is leading to the preparation of Shoreline Management Plans; and with the National Rivers Authority's programme for Catchment Management Plans, whose North West Region includes the Mersey Estuary in its 1996–97 plan preparation schedule.

THE MANAGEMENT PLAN'S POLICY PROPOSALS

Structure of the Plan

The study team's Initial Proposals (ULST, 1994a) for the Management Plan were put to the third annual Mersey Estuary conference and were subsequently subject to review by the Estuary Project Group during the spring and summer of 1994, with the Consultation Draft (ULST, 1994b) version of the Plan being published in October. Some 70 sets of comments, critiques and statements were received and analysed in the consultation process, which served to confirm the principal orientation of the Plan and to reshape and sharpen its structure. The principal orientation is a common one for the new perception of the waterfront in coastal area management identified by Vallega (1992), namely an overall approach to environmental resource use and protection, with 'sustainable development' being the key technical phrase.

The early structure of the Initial Proposals and Consultation Draft versions revealed some difficulty in expressing this orientation in an appropriately integrated form. The commercial navigation and port development issue had been one where environmental protection and economic development conflicts had been expressed (Alford, 1993) and indeed have been more widely recognised (Couper, 1992), but had been difficult to treat within the Plan's initial policy framework. The consultation process allowed for the issues to be worked through more deeply, using the concepts of the European Commission's 'sustainable mobility', and of environmental codes of practice (British Ports Federation, n.d.; ESPO, 1994), environmental management systems (Rennis, 1995), and environmental statements (MDHC, n.d.). The outcome was to provide the principal environmental protection policy area within the Plan with an over-arching function, setting the context for the other policy areas and expressing a number of key policies fundamental to the whole strategy (Figure 3.2).

In formal terms the Management Plan consists of three main hierarchically organised components: a 'Vision Statement'; a strategy made up of four policy areas corresponding to the four main groups of identified issues and conflicts and ten strategic objectives; and detailed developments of the ten objectives into policies and more specific, supporting management measures. Actions required in the implementation of the Management Plan's policies have been set out in a Policy Implementation Schedule indicating the main agencies likely to be involved, the intended time scale and, where appropriate, possible sources of

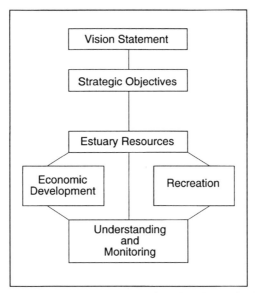

Figure 3.2. The Mersey Estuary Management Plan: the relationship between the 'Vision Statement', the estuary resources policy area and other policy areas.

funding. The Plan also contains two Appendices demonstrating how the Plan could be applied in practice to assess three hypothetical development schemes and indicating how certain policies may be viewed as attempts to resolve characteristic conflicts.

The Vision Statement sets out the main aspirations of the Management Plan and, while stated in part in journalistic phrases rather than technical language, none the less establishes the centrality of the new principal policy orientation and relates it to the subjects of concern to the commissioning body in a concise and robust way. The relationship between the Vision Statement and the main strategic policy areas and objectives is set out in Figure 3.3. The discussion here focuses on the Estuary Resources and the Commercial Navigation and Port Development strategic areas.

Estuary Resources

The Plan's primary concerns are set out in the introduction to the group of four Estuary Resources strategic policies: 'The overall aim of the Mersey Estuary Management Plan is to promote the sustainable use of the Estuary by ensuring that development of its economic and recreational potential is undertaken in a manner compatible with the protection, management and enhancement of the natural and man-made resources of the Estuary Zone' (ULST, 1995, 11). The Estuary Dynamics objective seeks to balance human interventions with natural conditions and processes, by proposing that the estuary be allowed '... to

VISION STATEMENT	STRATEGIC POLICY AREAS	STRATEGIC OBJECTIVES
'The Management Plan is based on a vision of the future of the Estuary as one of the cleanest developed estuaries in Europe, where the natural quality and dynamics of the natural environment are recognised and respected and are matched by a high quality built environment...'.	**Estuary resources** 1 Estuary dynamics	To allow the Estuary to function as naturally as possible and in a self-sustaining way by controlling human interference in inter-tidal and marine areas having regard to the natural conditions and processes of the Estuary and Liverpool Bay.
	2 Water quality and pollution control	To support continuing improvements in water, air, land, noise and light quality and the adoption of environmental good practice within the Estuary Zone.
	3 Biodiversity	To conserve, and where relevant restore, the natural biodiversity of the Estuary Zone.
	4 Land use and development	To promote careful stewardship of land resources, landscape and townscape within the Estuary Zone.
'a vibrant maritime economy...'.	**Economic development** 5 Commercial navigation and port development	To support the continued commercial and economic development of the Estuary's ports and port-related employment areas compatible with the Management Plan's environmental policies.
	6 Urban regeneration	To promote the regeneration of the Estuary Zone through maintaining and realising the distinctive potential of its existing developed waterfront and bankside locations and in adjoining areas.
'and an impressive portfolio of estuary-related tourism...'.	7 Tourism	To realise the potential of the Estuary as a focus for tourism.
'sport and recreation facilities'	**Recreation** 8 Informal and shore-based recreation	To maintain, enhance and, where appropriate, extend public access to the shores of the Estuary so that people may enjoy informal sport and recreation in safety.
	9 Water-based sport and recreation	To protect existing water-based recreation facilities and promote the appropriate development of new opportunities on the Estuary.
'The Plan will provide a framework for coordinated action. It will be a key instrument in addressing critical management issues so as to secure the sustainable development of the Estuary and to maintain and develop its position as one of the region's most valued environmental assets.'	**Implementation** 10 Understanding and monitoring	Steps should be taken to provide adequate management information to develop understanding and awareness of the natural dynamics of the Estuary and the inter-relation of social and economic activity - including the objectives and policies of the management planning process - with these natural factors.

Figure 3.3. The Mersey Estuary Management Plan: the relationship between the 'Vision Statement', the strategic policy areas and the ten strategic objectives.

function as naturally as possible and in a self-sustaining way ...' This is particularly important in maintaining the freedom of the river's low water channel in the Inner Estuary, re-established after maintenance dredging in the Eastham Channel was severely curtailed after 1962, which Hydraulics Research Ltd believe has allowed 'the Mersey to return to a state of equilibrium and that a balance has been achieved in the River's tidal processes, bringing to an end the net accumulation of sediment' (Cheshire County Council, 1989).

Specific Estuary Dynamics policies call for the Estuary Plan's concerns to be taken into account when assessing development proposals and existing adverse changes; and call for the restriction of new developments which would have the potential adversely to affect the semi-natural functioning of the estuary (including areas liable to flood risk) and/or to prejudice the capacity of the coast to form a natural sea defence. Other policies are directed towards coastal protection and flood defence activities, the compatibility of Management Plan policies with shoreline management and coastal defence in the Liverpool Bay Coastal Cell.

The Water Quality and Pollution Control objective endorses the water quality improvement objective of the Mersey Basin Campaign and extends it to cover air, land, noise and light quality and environmental good practice. It brings forward proposals for the relevant organisations to work together to establish and meet agreed pollution reduction targets and to see that new development does not adversely affect their achievement. A survey and review of contaminated land is proposed together with a coordinated programme of care and remedial treatment. Recognising the new status achieved by the estuary's wildlife and the potential impact of further improvements in water quality, the Biodiversity objective endorses the conservation and, where relevant, the restoration of the estuary's natural biodiversity. Policies are set out for site and habitat protection in terms of international, national and local statuses, while management measures set out the desirability of safeguarding measures and site-specific management plans. This policy provides a powerful endorsement of the much-awaited official designation of the Inner Estuary Site of Special Scientific Interest as having internationally recognised RAMSAR and Special Protection Area status. Other policies would provide protection for significant species present in the Estuary Zone, and promote the adoption of creative nature conservation in relation to new and existing developments, estate management and farming. The Land Use and Development policy section takes 'careful stewardship' as its expression of the sustainable development principal orientation, covering land resources, landscape, and, mindful of the archaeological, architectural and historic waterfront heritage in the areas, its townscape. In land resource terms new development is to be discouraged from the remaining areas of open coast and encouraged to locate in its existing developed areas. Two management measures suggest that local planning authorities should take a coordinated approach to identifying a coastal/estuary/river valley corridor or zone when reviewing their statutory and informal plans (e.g. Warrington Borough Council, 1995).

Commercial Navigation and Port Development

This policy section begins by establishing the Estuary Zone's seaport system in its local economic context and in its international role as a gateway for deep-sea and mid-sea routes linking with British and continental European hinterlands. Two further main roles are established: as a main port for Irish Sea ferry routes (now concentrated in the MDHC's facilities) and as a point of access for port-related industries (e.g. along the Manchester Ship Canal's Lower Reaches from Runcorn to Ellesmere Port). The estuary's smaller ports (e.g. ABP Garston), wharves and canalside docks provide complementary facilities to those of the larger operators. The three issues considered to require attention in the Management Plan were: the central theme of supporting the continued prosperity of the estuary's ports and their related industrial and service employment in ways compatible with the Plan's environmental protection policies; calling for attention to be given to concerns about the future of dredging spoil deposits (an issue of particular interest to the Manchester Ship Canal); and working out agreements for more widely acceptable mechanisms and levels of evaluation for the treatment of environmental as well as economic interests in development decisions.

Specific policies were produced to protect the existing commercial navigation channels in the Estuary Zone, subject to economic and environmental constraints, to require that new developments should take account of the need for continued access by shipping to ports using the channels, and recommending an early identification of port expansion proposals with a view to protecting their status in the interim. Other policies seek to identify port-related employment areas, including environmental upgrading; the availability of sites from the 'flagship' scale (as at Seaforth) to smaller, in-fill possibilities; and for enterprise and training support. Policies are proposed for a technical assessment of future dredging obligations and disposal considerations and for an Estuary Environmental Code of Practice, designed to address the problems of the ports' permitted development rights through a self-regulation process within a context of goals and objectives shared with the wider community of public and local authorities, residents and specialist interests. Related policies in the urban regeneration area express support for offshore oil and gas exploration and development in the Irish Sea and for accommodating their onshore requirements; and for the continuation of marine aggregate winning at current levels in the Estuary Zone, subject in both cases to the protection of existing navigation channels and the natural environment of the estuary.

CONCLUSIONS

This chapter has reviewed the experience of the Mersey Estuary's seaport and cityport systems in relation to their wider coastal area/zone and to changes in the recent external and operating environments of the functions which make up the activities of a seaport–cityport–coastal area. It has explored the implications of a significant change in local ecological values and desired water quality for the future of the estuary as well as the more well-known seaport redevelopment and waterfront regeneration initiatives. The perceived threats of major infrastructure projects have brought about a coalition of interests, which has taken forward the development of a new management interface between the cityport and coastal zone through commissioning an estuary management plan. With the publication of the Management Plan in 1995, the focus moves on to implementation and management processes.

Vallega (1992, 116) has spelled out the general need at this stage in the evolution of the waterfront for (i) information systems, (ii) control functions and (iii) decision-making institutions and organisations, specifically or widely concerned with coastal area management. He sees the final objective as being 'to create a management pole, consistent with the regional role that the coastal areas are acquiring.' The institutional structure of the Mersey Estuary, however, is fragmented and complex; it does not have a specific nor a wider-range organisation concerned with estuary management which could form such a management pole. If the Management Plan is to be taken up as a point of initiation, the challenge now lies with the coalition of interests which came together to support the preparation of the Plan to create such a 'management pole' or focus through open, inter-agency institutional structures which provide for understanding, participation and commitment to a continuing and evolving process.

REFERENCES

ACRM (Acting Conservator of the River Mersey) (1986), *Annual report* (London).

Adcock, B. (1984), 'Regenerating Merseyside docklands: the Merseyside Development Corporation 1981–1984', *Town Planning Review 55*, 265–89.

Alexander, B. (1982), 'Future improvements to the Mersey Estuary', *Journal of the Royal Society of Health* 102, 211–17.

Alford, J. (1993), 'The developed coast: the Port of Liverpool within Sefton: a development control view' in Jones, C. (ed.) *Planning for the coastal zone: conference report, Birmingham 1992* (Maidstone: National Coasts and Estuaries Advisory Group), 33–6.

Aspinall, P.J. and Hudson, D.M. (1982), *Ellesmere Port: the making of an industrial borough* (Borough of Ellesmere Port: Ellesmere Port).

Batey, P.W.J. and Kidd, S. (1993), 'The developed coast: the case of the Mersey Estuary', in Jones, C. (ed.) *Planning for the coastal zone: conference report, Birmingham 1992* (Maidstone: National Coasts and Estuaries Advisory Group), 27–32.

Bird, J.H. (1963), *The major seaports of the United Kingdom* (London: Hutchinson).

British Ports Federation (n.d.), *Environmental code of practice* (London: British Ports Federation).

Brown, P.J.B. (1990), *The development of Liverpool Airport: imaginative leap or flight of fantasy?* (Liverpool: University of Liverpool, Civic Design Working Paper No. 38).

Buckley, A.D. (1980), 'The Mersey estuary: a way ahead?', *Chemistry and Industry*, 19 April 1980, 321–7.

Chartres, J. (1980), 'Tidal power and airport ideas to boost Merseyside', *The Times*, 22 November.

Cheshire County Council (1989), *The ecology of the Mersey Estuary* (Chester: Cheshire County Council, Environmental Planning Service).

Couper, A.D. (1992), 'Environmental port management', *Maritime Policy and Management* 19, 165–70.

Dee Estuary Forum (1994), *The Dee Estuary Strategy: vision statement and goals report* (Birkenhead: Dee Estuary Forum).

Department of the Environment (1992a), *The Government's response to the second report from the House of Commons Select Committee on the Environment: Coastal zone protection and planning* (Cm 2011) (London: HMSO).

Department of the Environment (1992b), *Planning policy guidance: coastal planning* (PPG20) (London: Department of the Environment/Welsh Office).

Dixon, A. (1985), 'The Mersey Pollution Alleviation Scheme', *Journal of the Institute of Water Engineering and Science* 39, 401–13.

East Sussex County Planning Department (1992), *Draft strategy for the East Sussex Coast* (Lewes: East Sussex County Council).

EAU (Environmental Advisory Unit) (1991), *The Mersey oil spill project, 1989–90* (Liverpool: Mersey Oil Spill Advisory Group).

English Nature (1992), *Caring for England's estuaries: an agenda for action* (Peterborough: English Nature).

English Nature (1993), *Strategy for the sustainable use of England's estuaries* (Peterborough: English Nature).

ESPO (European Sea Ports Organisation) (1994), *Environmental code of practice* (Brussels, ESPO).

Gilman, S. and Burn, S. (1982), 'Docklands activities: technology and change', in Gould, W.T.S. and Hodgkiss, A.G. (eds) *The resources of Merseyside* (Liverpool: Liverpool University Press), 27–39.

Handley, J.F. (1992), 'Towards a management plan for the Mersey Estuary', in *Mersey Estuary Management Plan: Briefing Papers* (Liverpool: Estuary Project Group, Mersey Basin Campaign).

Heseltine, M. (1982), *Cleaning up the Mersey – a task for a generation*, Department of the Environment press release NW 179/82 (22 November).

House of Commons Environment Committee (1992), *Coastal zone protection and planning (2 vols) (Second report, Session 1991–92, HCP 17, 1–2)* (London: HMSO).

Hoyle, B.S., Pinder, D.A. and Husain, M.S. (eds) (1988), *Revitalising the waterfront: international dimensions of dockland redevelopment* (London: Belhaven).

Hyde, F.E. (1971), *Liverpool and the Mersey: an economic history of a port, 1700–1970* (Newton Abbot: David & Charles).

King, G. and Bridge, L. (1994), *Directory of coastal planning and management initiatives in England* (Maidstone: National Coasts and Estuaries Advisory Group).

Lynch, A. (1994), *Weathering the storm: The Mersey Docks financial crisis, 1970–74* (Liverpool: Liverpool University Press).

MBC (Mersey Basin Campaign) (1986), *New life for the North West* (Manchester: Department of the Environment).

MDC (Merseyside Development Corporation) (1981), *Initial development strategy* (Liverpool: MDC).

MDC (Merseyside Development Corporation) (1990), *Development strategy* (Liverpool: MDC).

MDHC (Mersey Docks and Harbour Company) (n.d.), *Environmental statement* (Liverpool: MDHC).

MDHC (Mersey Docks and Harbour Company) (various), *Annual reports and accounts*.

MDHC (Mersey Docks and Harbour Company) (1993/94), *The Port of Liverpool: handbook and directory* (Liverpool: MDHC).

Mountfield, S. (1965), *Western gateway: a history of the Mersey Docks and Harbour Board* (Liverpool: Liverpool University Press).

MSCSC (Manchester Ship Canal Steering Committee) (1985), *Manchester Ship Canal – Upper Reaches planning study: policies for the future* (Manchester: MSCSC).

National Rivers Authority (1995), *The Mersey Estuary: a report on environmental quality* (Water Quality Series No. 23) (London: HMSO).

NWWA (North West Water Authority) (1980), *A cleaner Mersey – the way forward* (Warrington: NWWA).

OECD (Organisation for Economic Cooperation and Development) (1993), *Coastal zone management: integrated policies* (Paris: OECD).

Porter, E. (1973), *Pollution in four industrialised estuaries: studies in relation to changes in population and industrial development* (London: HMSO).

Rennis, D. (1995), 'Environmental management systems for ports', unpublished paper presented to the Marine Environmental Management Workshop, London, 1995.

Rice, K.A. and Putwain, P.D. (1987), *The Dee and Mersey Estuaries: environmental background* (Liverpool: Shell UK Ltd).

Ritchie-Noakes, N. (1984), *Liverpool's historic waterfront: the world's first mercantile dock system* (RCHM Supplementary Series: 7) (London: HMSO).

SERPLAN (1993), *Coastal planning guidelines for the South East* (London: SERPLAN).

Tavare, J. (1986), 'Introduction' in *Mersey Basin Campaign* (MBC, 1986), 2.

ULST (University of Liverpool Study Team) (1994a), *Mersey Estuary Management Plan: second year report: initial proposals* (Liverpool: Estuary Project Group, Mersey Basin Campaign).

ULST (University of Liverpool Study Team) (1994b), *Mersey Estuary Management Plan: consultation draft* (Liverpool: Estuary Project Group, Mersey Basin Campaign).

ULST (University of Liverpool Study Team) (1995), *Mersey Estuary Management Plan: third year report: final draft of the plan* (Liverpool: Estuary Project Group, Mersey Basin Campaign).

Vallega, A. (1992), *The changing waterfront in coastal area management* (Milan: Franco Angeli).

Warren, K. (1980), *Chemical foundations: the alkali industry in Britain to 1926* (Oxford: Clarendon Press).

Warrington Borough Council (1995), *A nature conservation strategy for Warrington* (Warrington: Warrington Borough Council).

Woodcock, G. and Keen, E. (1994), *The implementation of an estuary management plan: organisational structures and institutional arrangements* (Liverpool: University of Liverpool).

4 Cityport Development and Regional Change: Lessons from the Clyde

ANDREW H. DAWSON
Department of Geography, University of St Andrews, UK

'It is time to focus attention not on the port/city interface as such but on the cityport/coastal zone interface, and to look at the relationships between port cities and the management of coastal zones.' This statement, which introduced the idea of a second British–Italian Seminar to readers of the Institute of British Geographers' *Newsletter*, could not have been written by a Scot. People in Scotland may speak of the city of Glasgow or, for that matter, Dumbarton, Gourock or Largs, but, in matters to do with seaborne trade, with ports and with port industries, their discourse is all with reference to 'the Clyde' (Figure 4.1). There have been, and remain, many individual ports on the Clyde, some of which have existed for centuries, and there has been great rivalry between some of them in the past, most notably between Glasgow and Greenock. But, ever since 1662, when Port Glasgow was established as the outport for Glasgow, the economy of West Central Scotland has been both focused on, and integrated by, a more extensive coastal zone than that around any one of the river's ports. Developments in one part of the region have been intimately related with those in others and, in many of the most formative cases for its economy and settlement pattern, have been connected with the means of transport which the Clyde – river, estuary and Firth – has offered. Although it took until 1966 to establish the unitary system of port management, stretching from Glasgow to the Cumbraes, which had first been mooted in the late 19th century – the Clyde Port Authority was the first of the estuarial bodies which the Rochdale Committee had proposed for the British port industry to be established – the Clyde has provided an example of the changing relationship between economy, settlement and port development at the regional, as well as the urban, scale for three centuries. It has also provided lessons as to what may happen elsewhere on cityport/coastal zone interfaces, and how such changes might be managed. This chapter outlines the general pattern of the rise and decline of the river's ports, three other types of development which have occurred in association with the port settlements in the coastal zone, and the story of the Hunterston peninsula, before drawing some theoretical and policy conclusions.

Cityports, Coastal Zones and Regional Change. Edited by Brian Hoyle.
© 1996 John Wiley & Sons Ltd.

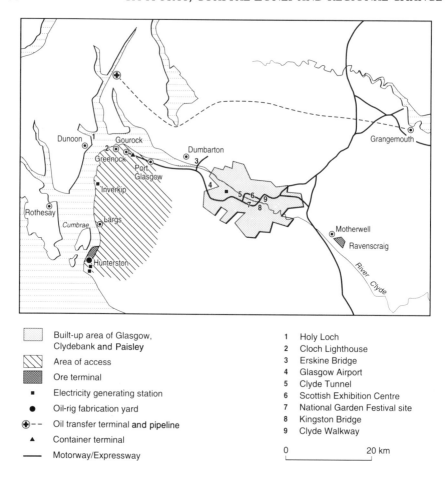

Built-up area of Glasgow,
Clydebank and Paisley

Area of access

Ore terminal

■   Electricity generating station

●   Oil-rig fabrication yard

⊕-- Oil transfer terminal and pipeline

▲   Container terminal

—— Motorway/Expressway

1   Holy Loch
2   Cloch Lighthouse
3   Erskine Bridge
4   Glasgow Airport
5   Clyde Tunnel
6   Scottish Exhibition Centre
7   National Garden Festival site
8   Kingston Bridge
9   Clyde Walkway

0   20 km

Figure 4.1. The Clyde

PORT DEVELOPMENT

The story of the Clyde is a classic of economic growth. Beginning with the transatlantic tobacco trade of the 18th century, and continuing with the construction of the Monkland and Forth and Clyde Canals (to bring coal from north Lanarkshire to the river), the Clyde valley has been the chief site of the modern cotton, metallurgy and shipbuilding industries of Scotland, the most populous area of the country, the location of its principal city, and the focus of its maritime trade. It has, however, also been a story of decay. Changing technologies, which have ushered in new means of transport and permitted the rise of new forms of industry, have gone on to destroy those same activities, replacing them with others, and in so doing have altered the relationship between the Clydeside towns and the river. Ocean-going ships no longer moor at the Broomielaw to unload cargo for the great warehouses in central Glasgow or to disembark migrants from rural Ireland to industrial Scotland; all the basins which were built on the upper Clyde between 1867 and 1931 have been closed; and many sites, from Motherwell in the central Clyde valley to Hunterston on the lower Firth, which were formerly used by industries associated with the river and its trade, lie idle. Passenger traffic is restricted to ferries across the Firth, and that of cargo, other than oil movements through the Loch Long terminal, is tiny in comparison with the levels of the early 20th century. Even the lower Clyde container terminal, built at what was to be the west coast terminus of the British motorway network in the late 1960s, ran out of customers in the 1980s, as the United Kingdom's trade turned increasingly from the North Atlantic and the Commonwealth towards Europe, and from goods to services.

Thus it is that the Clyde has experienced a wide range of those developments which might be included in any inductive model of port development. The initial construction of waterside quays and wharves – at which a wide variety of goods was handled – with their adjacent bonds, warehouses, processing industries and land-transport termini, was followed by the building of, first, tidal basins and, later, terminals dedicated to specific types of cargo. Meanwhile, increasing sizes of ship during this century have encouraged the development of port facilities further downstream, on the lower Clyde and at Hunterston, and led to the closure of smaller, older facilities alongside shallower water, further up the river, especially in Glasgow. In other words, individual ports along the river have been affected by these changes in different ways, depending on their location with regard to such matters as water depth and distance from the open sea; but changes in one cannot be understood without reference to those in others, and while in some circumstances it may therefore be appropriate to think of the Clyde ports as a series of distinct facilities, each related to an individual settlement, the changes which have been occurring in any one of them can only be explained in relation to those in its coastal zone as a whole.

OTHER DEVELOPMENTS IN THE COASTAL ZONE

The relationship between settlement and river along the Clyde has not, however, only been about those activities which have grown, on the strength of water transport, to be some of the largest in the world, only to disappear again within a few decades. Nor has it been solely about the changing technology of that transport and its effect upon the location of the industries and towns to which those activities have given rise. The establishment of not one but a whole series of settlements along the tideway has created at least three other types of demand for the coastal zone – for the generation of power, for recreation and for local transport – thus indicating the broad range of relationships between urban settlements and the coastal zone, and further illustrating the way in which the Clyde has been the focus of the regional economy of West Central Scotland.

Thus, the Clyde is central to any account of the changing location of electricity generation in the area. Following the establishment of a publicly owned electricity generating corporation in the late 1940s to supply central and southern Scotland (the South of Scotland Electricity Board), power generation ceased to be the responsibility of municipalities, and the search for power-station sites extended beyond the boundaries of individual local authorities. Furthermore, the growing scale of generating plant meant that only a few such stations were required. Estuarial and coastal sites were attractive locations in so far as they removed the need to build cooling towers, and several stations were built between central Glasgow and the Hunterston peninsula. A coal-fired station was constructed at Braehead in the late 1940s, an oil-fired plant was built at Inverkip in the early 1970s, and two nuclear stations were located at Hunterston – a Magnox plant in the 1950s, and a water-cooled reactor in the 1970s. The coal-fired plant has since become obsolete and been demolished. The oil-fired station, which was intended to use fuel imported to the Clyde Estuary, was completed shortly after OPEC quadrupled the price of petroleum, and has spent much of its life as a peak-load, rather than a base-load, generator; and the nuclear stations were sited at Hunterston, not only to make use of seawater for cooling, but in order that they should be at a 'safe' distance from the major centres of population and industry in West Central Scotland which they were to serve.

The Clyde has also been an important focus of the Scottish tourist industry, and the claim that 'The most scenic walk in the world is from Gourock Pier to the Cloch Lighthouse, and the second is that from the Cloch Lighthouse to Gourock Pier' illustrates the recreational importance of the coast for the people of the Clyde valley. Going 'doon the watter', usually by paddle-steamer, was one of the first forms of tourism for many of those living in Glasgow, Paisley and the other Clydeside towns in the 19th century, and dates from the sailing of Bell's steamship, *Comet*, from the Broomielaw to Greenock in 1812. Such an excursion – often only for a day – took people from central Glasgow to the sea lochs of the Clyde Estuary, while those who could afford longer, residential holidays stayed in hotels and guest houses in Dunoon, Largs or Rothesay. Very rich people built Italianate mansions amid large gardens in these and other resorts, overlooking the Clyde and offering views of the hills of Argyll or Arran, and travelled daily by ferry and train to work in Glasgow. Many others, with fewer resources, sought to emulate them. In 1946, the Clyde Valley Regional Plan identified what it considered to be a serious problem along the coast, namely, the unauthorised construction of chalets and temporary homes, and proposed that the area should be protected from such development. It did, however, suggest that the region should be opened up to the public for recreation, and that a large 'area of access', not only along the coast but also in the Renfrew Hills behind it, should be set aside in which such a use should have preference over agriculture.

In the event, the proposal was not taken up in this form and, as the tourist industry on the Clyde has subsequently changed and dwindled, as the Mediterranean and Florida have replaced Dunoon and Largs as principal venues for Scots' holidays, the coastal zone has experienced a very different fate. The three extant electricity generating stations mentioned above, an iron-ore terminal and reduction plant and an oil-rig construction yard have been built in the area which was to have been protected, and, for much of the period, a nuclear submarine base at the Holy Loch was added to the view from parts of it. It is, perhaps, ironic that it has been rather on the upper Clyde that tourism, in its broadest definition, has been actively promoted in recent years by central and local government, as a vehicle for the redevelopment of some of the docks and quays which have become redundant there, in the form of the Scottish Exhibition Centre, the Clyde Walkway and the National Garden Festival of 1988. It should be noted, however, that most of the Garden Festival site, which cost the government £35m to lay out at the former Princes Dock, quickly fell back into disuse and dereliction after the Festival was over, while the Exhibition Centre – built at a cost of £36m to public funds, £12m of which was for the reclamation of the former Queen's Dock – and Walkway attract much less custom than in their former uses.

The third type of demand which has been made upon the coastal zone as a consequence of the development of ports and their associated settlements is that of local transport. There has been, of course, a demand for transport facilities arising from the port activities themselves, but there are also demands which are related to other economic and social links between the towns along the river. Many of these were met by water in the early 19th century, but railways became increasingly important thereafter, and have been supplemented, and in part replaced, in this century by roads. Prior to the 1950s, George V Bridge in central Glasgow provided the lowest bridging point on the river, but growing demands for greater mobility have led to the construction of, first, the Clyde Tunnel between Govan and Partick, and later the Kingston and Erskine bridges – the latter of which lies 15 kilometres downstream from the city centre. There has also been an elaboration of the road network on both banks of the river since the 1960s, creating a motorway/expressway network which links Glasgow and areas to the north, east and south of the city with the north and south banks of the lower Clyde and with the new river crossings. It should be noted, however, that traffic on the Erskine Bridge has been so light that the income from tolls has failed to cover even the servicing of the debt which was incurred in the bridge's construction, and that use of the motorway between Glasgow Airport and the lower Clyde towns is also generally well below its capacity.

THE EXAMPLE OF HUNTERSTON

The changing relationship between economy, settlement and port development along the Clyde in recent decades is nowhere better illustrated than on the Hunterston peninsula. An extensive raised beach, covered with fertile soils, and formerly renowned largely for its early and seed potatoes, but with poor transport links with the rest of the Clyde valley, had been identified in 1946 in the

Clyde Valley Regional Plan as an appropriate location for one of two new holiday towns on the Ayr–Renfrewshire coast, 'set on one of the finest sections of the whole coastline' (p. 144). That town was not built, and within a few years quite different uses had been found for the area. In the mid-1950s, it was chosen as the site for a Magnox nuclear power station. It then became the focus of attention of the steel industry. Rapidly increasing sizes of bulk freighters, especially for iron ore, had underlined the wisdom of a 1929 study of the Scottish industry, which had suggested that the many small mills in north Lanarkshire (which had been founded on local, but declining, deposits of coal and ore) should be replaced by a large, integrated plant, immediately downstream of Glasgow, to which imported ore might have been brought by ship. That proposal was not adopted, and the proposed site was used for other purposes, but in the 1970s British Steel reclaimed a large area of foreshore, which the 1946 Plan had earmarked for a marina, at Hunterston and constructed a deep-water terminal and storage yard there, at a cost of about £100m. It also built an ore-reduction

Figure 4.2. A general view of the Hunterston Terminal, looking north up the Firth of Clyde. The deep-water jetty, with its cranes, stock yard, iron ore reduction plant and conveyor belt to the rail link with the former Ravenscraig steel works, are all visible. The jetty and stock yard, laid out on land reclaimed from the sea, are hidden from the coast road by an embankment, but are clearly visible from the holiday resorts of Fairlie and Largs, in the top left of the picture. (Reproduced with the permission of Clydeport and Scottish Nuclear)

Figure 4.3. The Hunterston 'A' nuclear power station, the first of the two to be built on the Hunterston peninsula, with the Firth of Clyde and the holiday resort of Cumbrae Island in the background. (Reproduced with the permission of Clydeport and Scottish Nuclear)

plant, in the hope that a major integrated steel works could eventually be established: in the meantime, it constructed a rail interchange in order to supply imported ore to the Ravenscraig plant at Motherwell, and closed its ore terminal at General Terminus Quay in central Glasgow, which was only capable of taking ships of 28 000 tons deadweight. Hunterston, with its flat land next to deep water, was also suggested in the 1970s as a site for an oil refinery and oil-rig construction yard, and simultaneously the South of Scotland Electricity Board was building a second, and much larger, nuclear power station adjacent to the first. In other words, there were many who anticipated that the Hunterston peninsula would provide a new coastal centre for the revival of several of those industries, and especially energy, metallurgy and heavy engineering/shipbuilding, which had proved to be so important to Scotland's industrial prowess in the past, and would create a growth pole, or, as some writers at that time might have regarded it, a Maritime Industrial Development Area (Hoyle and Pinder, 1981), in which 50 000 people would eventually be employed.

We now know, however, that almost all of the Hunterston developments were ill-conceived. The nuclear power stations will not cover their construction and decommissioning costs; the ore-reduction plants were never used; the Scottish steel industry, including the Ravenscraig mill, has all but closed, and the terminal has been sold; the rig yard is largely idle, and the oil refinery – the only one of the projects which would have been privately financed – was never built. Clydeport, the privatised port management company on the Clyde which purchased the ore terminal in 1993 for less than £5m, anticipates that potential custom amounts to only 12% of the terminal's bulk handling capacity. Rather than being a new industrial heart for Scotland, Hunterston has become a *fin de siècle* folly at the end of the Clyde's history as one of the world's major cityport/coastal zone industrial complexes, whose rusting industrial structures demean the view of the Firth of Clyde from several of its remaining tourist resorts.

CONCLUSIONS

Earlier in this chapter it was suggested that the story of the Clyde is but one, rather clear, case upon which the well-known, inductive models of seaport-cityport development might have been founded. The relationship between cityport and coastal zone may, however, be linked to a wider body of theory. Von Thünen, in the elaboration of his model of rural land use around an isolated city, introduced a line of improved transport, and indicated that land along it would probably be employed more intensively than would be the case in the absence of the *pro rata* reduction in transport costs which it would provide. He was thinking of a river, but the principle may apply equally to coastal sites. In other words, much of the relatively intense development downstream or along the coast from cityports would appear to be a reflection of the 'rents of situation' which such sites offer for the activities which are required by such settlements, as well as the 'rents of fertility' – broadly defined – which coastal sites that offer attractive scenery or water for industrial cooling may also enjoy. Mention of economic rents leads, however, to the recognition that the map of land use is not, *pace* Von Thünen, at equilibrium. Rather, it is subject to constant flux as economic rents – the surplus from economic activity – are appropriated by players in the space economy game and applied, often to different activities and/or in different locations from those in which they have been generated. Thus it is that the process of expropriation and reinvestment generates such structural economic changes as the Fisher–Clark sectoral shift and the changing modal split of transport demand, and such spatial effects as the transfer of older forms of manufacturing to low-cost economies and of tourism to sunnier and now more accessible climes – all of which have played a part in the changing relationship between cityport and coastal zone, not only on the Clyde but perhaps also elsewhere. It follows that, if it is the case that, like the Clyde, other formerly port-based regional economies have now developed as a result of such changes to the point at which the ports themselves play only a very small role, it may indeed be time to turn attention instead to the wider issue, which is the subject of this volume, of the cityport/coastal zone interface.

It is this, continuously dynamic character of the cityport/coastal zone relationship – which has been so clearly illustrated by developments on the Clyde in recent times – which suggests a number of lessons for those who would become involved in its management. First, it is clear that the demands that those who live in coastal cities make of the littoral are varied and changing. Second, it would seem that, to the extent that those demands are mediated through the collective wisdom of committees, especially those which are not responsible to shareholders but which can finance their schemes from taxation or the exploitation of their status as publicly owned monopolies, coastal developments may run a greater risk of being unnecessarily large, ill-timed in relation to the evolving nature of technical and economic change as a whole, and in the wrong place than might otherwise be the case. Third, the changing nature of those demands would seem to cast doubt on whether 'protection' – of either existing industries and facilities or of the coastal zone itself – can ever be justified. One man's subsidy, and one area's protection from change, may both be another's lost job opportunity. Finally, there is the problem of 'clearing up' those sites which have been used, and often left derelict, by activities which then cease to exist or move elsewhere. Perhaps those sites which have been, and remain in, public ownership should be leased – not sold – at whatever price the market will bear, subject only to the requirement that they should not be vacated thereafter in a worse condition than that in which they have been taken over, while owners of sites which are in private hands should be obliged to adopt a similar standard of behaviour.

REFERENCES AND FURTHER READING

Abercrombie, P. and Matthew, R. H. (1949), *The Clyde Valley regional plan 1946* (Edinburgh: HMSO).

Hoyle, B. S. and Pinder, D. A. (eds) (1981), *Cityport industrialization and regional development: spatial analysis and planning strategies* (Oxford: Pergamon).

Leighton, J. M. (c. 1830), *Swan's views of the watering places on the Clyde* (Glasgow).

Ministry of Transport (1962), *Report of the Committee of Inquiry into the major ports of Great Britain* (The Rochdale Committee) (London: HMSO, Cmnd 1824).

Patterson, A. J. S. (1969), *The golden years of the Clyde steamers* (Newton Abbot: David & Charles).

Patterson, A. J. S. (1972), *The Victorian summer of the Clyde steamers* (Newton Abbot: David & Charles).

Riddell, J. F. (1979), *Clyde navigation: a history of the development and deepening of the River Clyde* (Edinburgh: John Donald).

Smith, R. and Wannop, U. (1985), *Strategic planning in action: the impact of the Clyde Valley Regional Plan 1946–1982* (Aldershot: Gower).

Thünen, J. H. Von (1826), *Der Isolierte Staat* (Rostock).

Part II

ENVIRONMENTAL ISSUES

5 Environmental Perception and Planning: The Case of Plymouth's Waterfront

LUCIANO CAU

Department of Geography, University of Southampton, UK

The waterfront, that magic line where the elements of earth and sea meet and intermingle, has always both attracted and repulsed human beings. Archimedes' Law, stressing the utility of water as a medium for transport, has been the main attracting force, polarising on the coast a wide range of productive and commercial functions. Also, water sites have in many cultures a particular aesthetic value. At the same time, waterfronts have often been dangerous places where the forces of nature show their full destructive power. In the long run, as human capacity to subdue these forces grew, attraction largely overcame repulsion. Paradoxically, the same dramatic display of the elements' fury gives the coasts the sublime character so appreciated by Romanticism, making them even more attractive. However, the success of coastal zones in attracting people and activities, an attraction that has reached sometimes paroxysmal levels, is threatening their integrity.

In the late 1960s, two developments took place that changed attitudes in this respect: the application of General Systems Theory to the study of ecology, and the rise of the environmental movement. Researchers demonstrated that coastal zones are fragile and complex systems: incautious intrusion and overexploitation in some parts of the system would badly affect the equilibria elsewhere. Environmental concern spread to most industrialised countries and the desirability of industrial development, until then taken for granted, became increasingly questioned. The conflict became particularly acute in coastal zones where competition for different land uses was more intense. The need for careful coastal zone management became increasingly evident on environmental and political grounds.

A basic role is played in this context by cityports and their urban waterfronts as they represent 'the most important component of demographic and economic pressure influencing coastal ecosystems' (Lucia, 1994, 15). Recent technological change in maritime transport, namely the growing size of bulk cargo vessels and

Cityports, Coastal Zones and Regional Change. Edited by Brian Hoyle.
© 1996 John Wiley & Sons Ltd.

the consequent need for deeper water, larger areas for storage and better access to inland transport networks (a common problem shared by containerised and general cargo handling), forced trade and related activities out of historic ports. The old areas are not suitable any more for this new intermodal transport. 'The modern seaport acts as a gateway rather than as a central place' (Hoyle, 1988, 3). Also, the economic restructuring process has forced many productive activities to reduce their workforce and relocate in places with a lower level of environmental opposition. As a consequence, formerly strong economic and cultural functional links between port and city loosened or disappeared altogether (Hoyle, 1988). This has left sailortowns (Hilling, 1988) in a decadent state, with severe unemployment problems and bad living conditions.

Revitalisation of these urban spaces can be an important mechanism for improving environmental quality in these inner port cities and for achieving a better quality of urban life. However, most such plans are drawn up pretending that planners know what is better for cities as wholes and for the citizens alike. Plans are usually based on the aim of improving a certain number of specific physical or economic characteristics of the areas they are concerned with: to reduce overcrowding and clear up slums, to increase income levels through gentrification, and to diversify economic output involving the relocation of productive and commercial activities. Some environmental issues are taken into consideration, either to please vociferous environmental groups or because of the personal commitment of planners. However, it seems that it is at least to some extent true that 'traditionally planners have designed urban environments without consulting their user clients' (Pocock and Hudson, 1978). Until recently, people usually welcomed, at least apparently, development plans drawn up in this way. Interested communities did not normally discuss the suitability of these decisions. Nowadays, on the contrary, this happens very rarely, at least in the most developed countries. Environmental groups paved the way in this respect, and an era in which the influence of communities affected by the decision-making process has replaced one in which unquestioned public support for external decisions was largely assumed.

One could argue that the planning system is, in a democratic society, normally and naturally open to suggestions and criticism, and anyone who has a word to say is allowed to comment. However, this chapter suggests that if waterfront redevelopment aims to improve local environments especially for local residents, there is a need to explore what perceptions such residents have of their local environment. We can paraphrase a well-known statement and say that: 'There is no such thing as "the environment"'. As far as perception is concerned, individuals pick up different aspects of the environment and arrange them in different sets. This is likely to lead to a certain level of mismatch between public expectations and planners' proposals that at best leaves residents with a feeling of powerlessness and frustration, and at worst may lead to anger and open conflict situations. There is no need for this to happen and proposals are made here to widen the opportunities for public consultation. From the extensive waterfront of

Plymouth, the Cattedown peninsula has been chosen as a case study. The evolution of both environmental quality and local residents' perceptions and reactions is reconstructed through a sequence of historical maps and published accounts. Present-day perceptions of the local environment have been explored through a questionnaire survey personally administered. Planners' perceptions of local needs are derived from recently published local plans.

AN EXPLANATORY MODEL

This chapter is based on research originally derived from a model proposed by Pinder (1981, 1984; and in Clout *et al.*, 1985), who stresses the growing role that interest groups in society, such as planners, medical professionals, environmental groups and public opinion, generally play in influencing the development of the decision-making process. He suggests that 'legislation can, in fact, be conceptualised as a ceiling or "society tolerance limit" below which proposed industrial developments must stay if they are to be accepted' (Pinder, in Clout *et al.*, 1985, 1987). This legislation changes in a spatial dimension, as different countries differently enforce environmental issues. A flat surface (Figure 5.1) covering a whole country represents a uniform legal limit. Its level is fixed as a result of the democratic process of mediation between opposing pressures. Developments whose impact exceeds such limits are illegal. However, the limit itself could be too weak to meet community desires. Pinder also points out the role of environmental groups which, feeling legal limits to be too loose and to be therefore represented by an inappropriately high tolerance limit, pressure relevant authorities to lower the 'society tolerance limit' to a more acceptable level.

This model has been mainly designed to work at a national or regional scale. It is also based on the assumption that either legislation or environmental groups represent society as a whole. We can adapt and apply the model to the local level by considering a single individual's limit of tolerance, with possibly different levels for different individuals and varying evaluations of different features of the environment. In this way we can take into account that members of the same community may hold different views and also that the environment, in everyday life, means not only such things as air or water quality, but also more prosaic aspects such as litter collection, house façades or parking spaces. We can still consider a resultant surface as a representation of the distribution of average attitudes per unit of geographical area. This irregularly shaped surface may be termed a 'community tolerance limit' (CTL). A CTL changes in space and time, and the model still keeps a flat surface representing the legal tolerance limit (LTL), as expressed in law and local plans.

The wider the gap between the CTL and the LTL, the stronger the long-term pressure on the legal and planning system to put stricter limits on the use and abuse of the environment. At this stage we distinguish between objective or *actual* environmental impacts on one side (AEI) (i.e. every induced change in

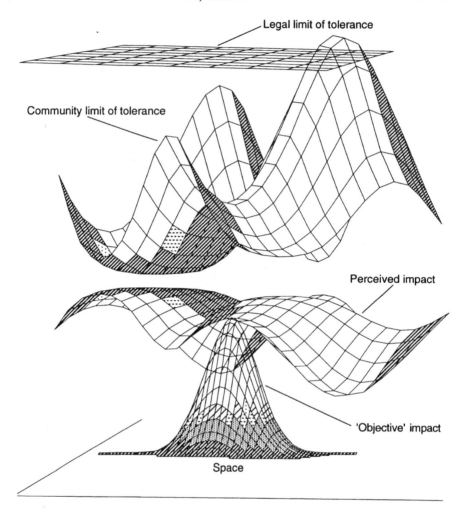

Figure 5.1. Explanatory model of variations of tolerance level to environmental impact. (after Pinder, D.A., in Clout, H. *et al.*, 1985)

biological, physical and chemical parameters of the environment) and the subjective or *perceived* environmental impact (PEI) on the other. Therefore, we should include in the model two other surfaces. The AEI, without external influences, will have the shape of a solid generated by the rotation of a Gaussian curve. The PEI instead will have a much more irregular shape, depending on many physical and socio-economic variables. The LTL deals with the AEI. Everything that is below the LTL is fine. However, people's reactions are based on their perception of such impacts. Sometimes the two surfaces, the PEI and the AEI, are not even related.

Most normal activities stay well below the CTL. However, the more such activities impinge on a community's well-being, its possibilities for recreation or on the natural environment, the higher the probability that the CTL will be reached and even exceeded. Every intersection of the PEI surface and the CTL is likely to precipitate a conflict situation. The deeper the intersection, the higher the probability of conflict and significant external opposition. In democratic countries, such conflicts can lead either to the closure of existing industrial plants or to the abandonment of or major alterations to projects, according to community desires. What this chapter aims to show is that what really matters in triggering conflicts is not the objective or actual impact (AEI), but the subjective or perceived one (PEI).

Another problem that we have to tackle is finding out what is the mechanism involved in shaping and changing attitudes towards the environment. A group of researchers from Clark University, USA, recently proposed a theory that tries to explain why some risks receive more attention by a society while others, sometimes even more dangerous, receive less (Royal Society Study Group, 1992). They suggest that danger and its objective characteristics interact with a number of psychological, social and cultural processes, in ways that intensify or attenuate the perception of risk. The core of this hypothesis is that we acquire most of our knowledge of the external world by receiving signs, signals and images through an ever-changing communication network. Nodes of such a network are represented by 'stations of amplification' (such as parents, relatives, neighbours, friends, but also teachers, mass media, environmental groups, etc.). The behaviour of each station (that is, the aspects they amplify or attenuate) is largely predictable according to social structure, circumstances and objectives. Therefore, if we know the influence these stations exercise on a social group, we can also guess with good approximation the group's perception of environmental change; and we can possibly predict the behaviour of the CTL.

ENVIRONMENTAL CHANGE IN CATTEDOWN, PLYMOUTH

It seemed worthwhile to try to unify the model and the theory outlined above. In fact, in many ways they complement each other. The resulting theoretical tool has been applied to Plymouth's waterfront in a study that includes an analysis of the evolution of attitudes toward the local environment shown by the residents in an old port area.

Plymouth's waterfront is a complex entity (Figure 5.2). The coastline extends for 11 miles (18 km), from the mouth of the River Tamar to the River Plym. 'It has established employment and residential areas but also includes the historic Hoe and Barbican; a port and a ferry terminal; a fishing harbour and Naval docklands; three conservation areas, two SSSIs, areas of outstanding natural beauty and two spectacular Ancient Monuments. There is also a country park, beaches, piers and marinas, not to mention a lake, a castle and an island!' (City of Plymouth, 1989, 16–17). This study focuses attention on one segment of this

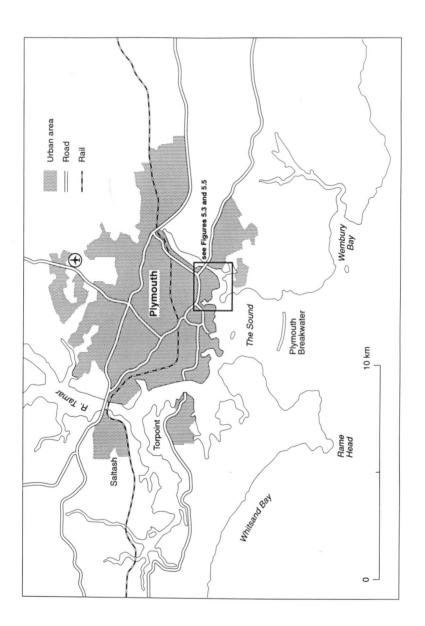

Figure 5.2. Plymouth and its waterfront

waterfront: the Cattedown peninsula (Figure 5.3). This includes the localities known locally as Prince Rock, Cattedown and Coxside, but for present purposes the entire area is included within the term Cattedown.

Cattedown lies at the eastern end of Plymouth's waterfront. Changes in the physical environment have been reconstructed through a sequence of historical maps, while the evolution of environmental perceptions has been mainly deduced from a number of articles in the local press. This particular area was chosen because it is an old industrial zone in which factories, warehouses and dwellings have developed together since the last century. By looking at its history, it is

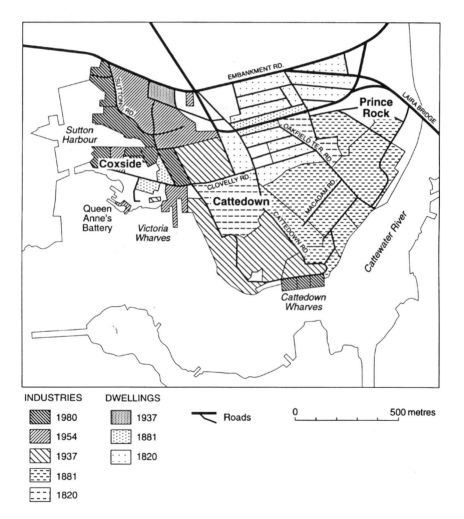

INDUSTRIES	DWELLINGS		
1980	1937	Roads	0 500 metres
1954	1881		
1937	1820		
1881			
1820			

Figure 5.3.   Cattedown, Plymouth: expansion of the built-up area reconstructed through a series of historical and modern maps.

possible to show how attitudes to environmental issues have changed over time. This in turn allows observations to be made concerning changing community tolerance limits over a long period. In addition, because this is a complex area of numerous land uses, in which urban decline is now well established, extensive planning has been carried out to redevelop the area. This area therefore provides a good opportunity to establish how far environmental perceptions and wishes of local people agree with official plans.

In 1643, a year in which Plymouth was involved in a siege during the Civil War, the Cattedown peninsula was open countryside. Sutton Harbour was then the only port of Plymouth, but quays and warehouses were located on the western side of the harbour in what is today known as the Barbican. By 1820 the area had evolved substantially: while most of the area was still devoted to agriculture, the waterfront had undergone major changes. Quarrying of the local limestone had opened wide scars in the southern and south-eastern parts of the peninsula; quays and piers started to diversify the shoreline and a group of warehouses had been built. A branch railway skirted the edge of the peninsula, serving the quarries, the ferry to Orestone and merchant traffic from sea-going vessels. Another branch railway ended at Sutton Harbour, still the busiest part of the commercial port. Productive activities had expanded eastwards, and a few ropewalks had been established in the area. Industrial and commercial land uses became predominant in the area by 1881. Gas works, tanneries, manure works and warehouses were the main elements in the urban landscape in Cattedown. Residential land use was also expanding and sometimes intermingled with industrial and commercial premises. Quarrying was developing northwards, leaving behind wide empty spaces surrounded by steep cliffs, spaces quickly occupied by new activities.

In the process of successful growth shown by the city of Plymouth as a whole around the turn of the century, the Cattedown peninsula was the only area of reasonable size available with waterfront access for the expansion of productive or commercial activities. On the western side of the headland on which Plymouth and the associated towns of Devonport and Stonehouse were expanding, the waterfront was occupied by or reserved for the Naval Dockyard. This has always been the major source of employment and prosperity for the city (and the main cause of destruction during the Second World War). By 1937, on the eve of the conflict, the economic success of the Cattedown waterfront had attracted a substantial population there. Reports in the *Western Evening Herald* explain that 'the landing of many varied cargoes meant a large number of dockers, many of whom lived in the area'. Among the goods landed were machinery, cement, ores, manure, timber, coal, petrol, and potatoes from Ireland. From Victoria Wharves steamers sailed to Aberdeen, Bristol, Cardiff, Dublin, Dundee, Hull and Liverpool. Those were the days when there was 'the atmosphere of a village within a city', but also when 'to "work the ships" in Cattedown meant very long hours, low pay and unbelievably hard conditions'. Houses were small and often unhealthy. Most dwellings lacked basic amenities and people had to bath in 'those galvanised baths in front of an open fire ... Canvas covered some floors, with distemper – usually pink – on the walls'. Road traffic was so limited that 'you could play alleys or marbles' in the streets. However, this did not mean that the environment was attractive: today we would not describe 'massive steam engines pulling fair wagons' as environmentally friendly, nor the traffic of 'huge shire horses' that left behind a lot of manure which, even if it was 'much sought-after' and 'cleared up quickly by local gardeners', must have contributed to an unpleasant environmental situation, by modern standards. A number of large coal heaps occupied most of the space around the gas works and coal dust was everywhere, not to mention the smoke from burning coal in fireplaces and stoves in the houses (*Western Evening Herald*, 19 and 26 September, 3 October 1985).

People's tolerance to such unpleasant conditions – the CTL of our model – must have been quite high. People living in Cattedown grew up there or in similar areas, and the same was true of their parents, relatives and neighbours. These communication networks, as defined above, probably did not consider such conditions as negative as we do today. Possibly they were accustomed to such environmental conditions and dust, dirt, smoke, noise and risk were considered just part of life. This attitude remained substantially unchanged until quite recently. Cattedown was badly damaged during the Second World War (Maguire *et al.*, 1987) but, unlike other parts of the city, its physical structure did not change with the reconstruction. Abercrombie's plan for Plymouth did not apply to this area. Space available was exclusively devoted to economic uses and the general layout of the built-up area has since been largely frozen as it was in the pre-war years. Certainly, the coal-fired power station and the abattoir with the adjoining meat market and glue factory, established in the area during the reconstruction, did not greatly diversify or enhance the local environment. According to some local evidence, butchery remnants were dumped straight into the river, with consequent 'awful' and 'horrible' smells spreading up to the dwellings.

The trade in oil products is of particular interest in the examination of the evolution of local attitudes to industrial and commercial activities in the area. Storage and distribution of oil products were well established in the area by the early 1920s. In the late 1960s, revenues from this trade represented almost 30% of the total receipts entering the port of Plymouth. Based on a range of products then relatively new, and with a 'high-tech' profile symbolising modernisation in many respects, this led to a £1.5m scheme to modernise the Shell-Mex and BP terminal at Cattedown Wharves, announced in 1969. The main work was the building of 11 new tanks between 60 and 70 ft high (22–26 metres) ('among the highest in use in this country!' the *Western Evening Herald* proclaimed) in an area of 5 acres (2 hectares). The project also included dredging the channel to the wharf and enlarging the turning area to grant access to tankers of up to 24 000 tonnes deadweight (dwt). Even if the project was not intended directly to generate new employment, it was believed that it 'could help attract industry to the city and at the same time play an important role in regaining some of Plymouth's former glory as a thriving commercial port' (*Western Morning News*,

4 June 1969). The company also intended to double its road tanker fleet to 83. In 1970 another oil company, Conoco Ltd, announced that they were going to install a new petroleum storage terminal in Cattedown, on a site previously occupied by a South Western Gas Board tar distillery.

It is noteworthy that in the 1960s people really welcomed such announcements and companies did not conceive that their progressive plans could meet opposition. Evidently the PEI was not only below the LTL but also below residents' CTL. People simply did not consider the environmental problems created by this kind of development. Job creation and economic growth were clearly the most important issues. This is a typical reflection of a situation in which community tolerance to environmental impact is very high. The possibility of unlimited development was seen to be threatened by no serious constraint other than the availability of the factors of production.

However, on a wider front, attitudes were relentlessly changing. Adaptation to locally inherited values was no longer the only factor that shaped people's perceptions. Mass media in many ways changed patterns of diffusion of innovations. They channelled not only information, but also new symbols and values. During the late 1960s, concern about the environment became a dominant issue in the national media. This was not just an artificially induced interest but was part of a wider cultural trend. Attitudes toward industrial development and its impact on the environment were dramatically changing in most of the developed countries (Pinder, 1981, 1984). Industrial development and the quality of life were no longer seen as going unquestionably in the same direction. This change is probably related to the evolution of the economic structure of societies, with the increased role of the service sector and the staggering increase of productivity in agriculture and manufacturing industry. It can also be seen partly as a reaction against some effects of this increased productivity. Cultural neo-Romanticism, with an idyllic view of nature, a rejection of what is modern and artificial, or urban, has probably also played a part in these processes.

In Cattedown, these changes did not become apparent until the late 1970s. Some local residents harshly complained of environmental problems created by the economic activities in the area. Between 1975 and 1979 protest was voiced and letters published by the local press. It was not until 1980, however, that newspapers turned attention to what was going on in this part of Plymouth. On 28 January 1980, the *Western Evening Herald* titled its front page: 'Heavy lorries that rumble around the roads of Cattedown are steadily destroying the environment'. The range of complaints voiced in the article were mainly targeted at businesses located in the area: pollution, heavy traffic, noise, vibration, intrusive car parking, etc. Lack of recreational spaces for the children was also one of the main problem points, partly because the South Western Electricity Board and Plymouth's waste disposal service had bought the few such spaces available and turned them into car parks for their employees.

Objectively, conditions were not much worse than a few years before. Apparently the CTL, reacting to disturbing activities, suddenly lowered and the

PEI was no longer below the threshold. As a result, potential conflict became apparent. The situation worsened when an old rail track was dismantled and became the present Gdynia Way, a trunk-road bypassing Embankment Road, to lighten the intense traffic along the latter. To keep the traffic flowing smoothly around the roundabout where these streets join, a footbridge was installed. Residents in Cattedown vigorously objected that the elderly and persons with reduced mobility found it extremely difficult to use this bridge. They presented a petition to introduce 'a properly controlled crossing for pedestrians' instead of the bridge, but they did not succeed (*Western Evening Herald*, 4 December 1980). A local shopkeeper, Mr H. Chaffe, was the backbone of the protest and he frequently appeared in the local press making local residents and the general public increasingly aware of the environmental problems in Cattedown.

However, another swing in the interest of the press occurred after 1985, stressing this time the economic potential of the area for job creation for the whole city. This change of viewpoint probably reflected general public concern due to the recession, particularly after the privatisation and restructuring of the naval dockyards made about 8000 workers redundant. This also underlines the important role the area plays in the economy of the city as a whole.

PORT AND PEOPLE: THE DECLINE OF A LONGSTANDING RELATIONSHIP

Because of its physical characteristics, both in terms of its geographical position in the context of the economic geography of the UK as a whole, and as a reflection of the morphology of the site itself, the Cattedown waterfront has been unable to meet the new requirements of modern maritime transport technology. Moreover, the Naval Dockyard has tended to have a monopolising influence over much of the non-naval waterfront. This might have limited the possibilities of endogenous development in the city and in the Cattedown section of the port in particular. It is rumoured that competition for highly skilled workers was a major concern for the Ministry of Defence in Plymouth. Apparently this induced the MOD itself to veto any proposal for major industrial development in the catchment area of the city.

Notwithstanding decline and redevelopment problems, Cattedown is still an important part of Plymouth's economy. This is the case although manufacturing has diminished both in productive and occupational terms in favour of commercial warehousing and services often not related to the presence of the port. Port facilities of this waterfront comprise Cattewater Harbour and Sutton Harbour (see Figure 5.3). Cattewater Harbour, formed by the mouth of the River Plym, has an area of about 105 hectares and can accommodate vessels up to 12 000 dwt. The main piers are Cattedown Wharves and Victoria Wharves, both provided with facilities for handling bulk and general cargo, but mainly the throughput is represented by oil products, china clay, grain and fertilisers. Sutton Harbour is the old port around which Plymouth thrived for centuries, at the western end of the Cattedown peninsula. The cargo handled there includes coal imports, steel exports and general cargo, but it is today more important as a marina (with 350 berths) and fishing port. In the south-eastern part of the harbour there is Queen Anne's Battery, a marina with 300 berths and a luxury yachting club. Space for small-craft mooring is available on both sides of the Cattewater as well. These uses are intermingled, not always happily, with a number of water-based leisure activities (bathing, surfing, canoeing, water-skiing, etc.). The harbour also receives sewage discharges, partly not previously treated. There are, therefore, important complexities and conflicts in this part of the waterfront and these problems arise partly from its historical legacy and partly from the pressure of new developments.

In 1990, only 77 of the 296 firms established in the area were engaged in port-related activities, or at least water-related; 48 of these were broadly connected to fishing and watersports and the other 29 were mainly oil products wholesalers (City of Plymouth, 1990). Of the remaining 219 firms not related to water or the waterfront, 24% were vehicle repair and maintenance services, 11% were related to the building industry and 4% catering and food processing. These are the categories with the highest frequency and this gives us an idea of the considerable range and variety of activities based in the area. The fact that port activities, although they are prominent land users in Cattedown, do not provide great economic impetus is well demonstrated by the oil depots. Bulk fuel traffic, most of which passes through the Cattedown terminals, amounted to 1 million tonnes in 1965 but after a peak of about 1.3 million tonnes suddenly collapsed during the oil crisis in 1973, so that in 1975 the throughput was no more than 750 000 tonnes. The level in 1990 was still lower than in 1965, with just 900 000 tonnes. Moreover, the traffic of oil products is 99.4% inwards. In this case the port and depots are just links in the delivery chain.

In urban terms this change in the relative importance of activities in the area can be seen as part of the normal economic and land-use evolution of the city. In local terms, however, it reminds us that the relationship between residents and local activities has changed dramatically. The old economic and cultural functional links that used to tie people to the wharves and to the ships they worked are gone, probably forever. It can be argued that this basic change lies behind many of the attitudes to economic activity and employment which the survey in Cattedown was able to identify.

Since its origin, between the later 19th century and the early 20th, the overall structure of the Cattedown residential area has remained virtually unchanged. Quarrying expanded from the south and housing from the north until they nearly met. With the industrial development of the area, the situation crystallised except along a narrow belt that marks the boundary between these land uses. Along this belt modest loss and re-conquest of space occurred. It is also the area where potential for conflict from environmental causes is higher, because of the sheer proximity of contrasted land use. The dwelling stock amounts to 1395 units, 43% of which are flats and 54% terraced houses. The quality of the building materials is quite poor and the general appearance is fairly uniform. Warehouses, depots and redundant industrial spaces contribute considerably to the generally neglected and bleak atmosphere. Some industrial estates are fairly well concealed in the bottom of craters left by quarrying, and some landscaping has been attempted, particularly in the premises of the oil companies. But in most cases no attention is paid to the visual appearance of the buildings and the aesthetic intrusion can be significant. This negatively affects property values. The houses occupy the lowest segments of the real estate market. They are hence generally owned by or rented to members of low-income social strata. The demographic structure of the area is quite irregular. Nearly 20% of the 2875 residents are aged 65 years or more; 29% are between 20 and 34 years old and nearly 8% are in the cohort of 0 to 4 year olds (City of Plymouth, 1992). In some ways these data account for some of the results of the survey.

THE SURVEY

Fieldwork was carried out in Cattedown in June and July 1992. It was based on a questionnaire survey, designed to be personally administered to 100 interviewees. This was done to gain insights into a certain number of aspects of the attitudes residents show towards the local environment. A systematic sample of households was designed, randomly selecting the first one in which an interview was to be administered and the next being one in every thirteen for each side of the street. A first version of the questionnaire was produced and tested in a pilot survey. This led to some rewording of several questions in a more colloquial language, to make them fully understood. Some questions proved to be irrelevant and were deleted before a final version (Table 5.1) was produced. A number of difficulties emerged. First, problems were experienced with vacant properties and ones where contact could not be made with the occupier. Second, it quickly became apparent that many people were suspicious of the questionnaire, thinking that it might simply be an excuse to gain information which might be useful in, for example, burglary. Almost certainly, part of the difficulties were due to the interviewer's Mediterranean appearance. It became evident that a systematic random sample was not going to work. In order to obtain a substantial sample, it therefore became necessary to try to contact a larger number of houses in the area. Also, it was necessary to carry out interviewing during the evenings and weekends, as well as during working days, to ensure a reasonably good cross-section cover of the adult population. This meant that in the time available it was possible successfully to complete only 80 interviews, but this sample represented a fairly good cross-section of the adult population, both by gender and by age, although the 30–49-year-old group was over-represented by about 12%.

Two other features of the sample need to be emphasised at this point. First, one-third of those interviewed had lived in the area for less than five years. Indeed, 18% had moved there within the past year. While the area has a substantial core of long-established residents, it is also one of high rates of

Table 5.1.   Selected questions from the Cattedown questionnaire

Question 1	How long have you lived in Cattedown?
Question 2	Have you ever experienced any problem or nuisance by living close to industrial plants in Cattedown?
Question 3	To what extent do you think your street is: (2 = Very; 1 = Moderately; 0 = Neither)

	2	1	0	1	2	
Safe	☐	☐	☐	☐	☐	Dangerous
Friendly	☐	☐	☐	☐	☐	Forbidding
Beautiful	☐	☐	☐	☐	☐	Ugly
Cheerful	☐	☐	☐	☐	☐	Gloomy
Healthy	☐	☐	☐	☐	☐	Unhealthy
Clean	☐	☐	☐	☐	☐	Dirty

Question 4	Do you have any ideas about how you would like to see the derelict land, here in Cattedown, used?
Question 5	Do you have any ideas about the parts of Cattedown in which these improvements ought to be carried out?
Question 6	Can you suggest any other ways of improving the local environment in Cattedown?

Other questions gathered information on the respondent's characteristics (age, sex, economic status, income, educational qualifications, marital status, dwelling type and tenure, work location, etc).

residential change. Second, 40% of those interviewed spent the majority of their time in the area and were therefore exposed almost continually to its environment. Most of these interviewees spent the majority of the day at home, but very few were employed in the area. Conversely, more than 50% of the sample was employed outside the area. This limited their experience of the local environment to the evenings and weekends.

Because the aim of the survey was to elicit attitudes, some of the questions were left open-ended. The general attitudes revealed were quite evenly divided. Of the 80 respondents interviewed, 42 declared they never experienced problems or nuisances by living close to industrial plants in Cattedown. The other 38 claimed the reverse. Not one of the surveyed variables clearly related to these differing attitudes. It was hypothesised that people spending most of the time in the area would experience the worst problems and therefore show stronger feelings, but this proved not to be the case. Also, the opposite was tested, i.e. people spending most of the time outside the area would feel disturbed by, for example, the traffic of heavy tankers during the night, because they were not accustomed to the continued exposure, but this was not confirmed. It was supposed that education could influence these attitudes, as environmental concern has become a main issue in education since the late 1960s. In this area, however, this expectation has not been satisfied. It was also believed that income could make a difference and it was hypothesised that relatively higher income could raise expectations in terms of quality of life, but this as well was not confirmed. However, most respondents felt the question on the household's income quite sensitive, therefore answers could be unreliable. Conversely, very few believed they were getting any advantage out of the closeness to warehouses and industrial plants. Just three positive answers were obtained, from people working in local firms. The conclusion drawn from these and similar results was that, even if the sample had been much larger, it would have been unlikely to have identified sub-groups in the population which held differing perceptions of the area. For this reason, further analysis concentrated on the sample as a whole.

PERCEIVED ENVIRONMENTAL QUALITY

It was thought that a better understanding of attitudes, i.e. how different aspects of the environment were evaluated, could be gained by asking about different aspects of the directly experienced environment, the respondent's street. Six different environmental attributes were selected for investigation, referring to physical, aesthetic, social and psychological characteristics, namely: safety, friendliness, attractiveness, cheerfulness, healthiness and cleanliness. These aspects constituted question three, designed as a set of six pairs of polar opposing attributes and organised in a five-step semantic differential scale. Judgements could take the value of '0' indicating an indifferent view on either attribute of each pair; '1' and '2' respectively added the reinforcing adverbs 'moderately' and 'very' to the attribute.

In this case also, not one of the relationships that were hypothesised was confirmed by the analysis. There is no evidence to confirm the idea that age and/ or gender explain views on safety in the area. For example, regression analyses for men and women produce coefficients for the independent variable that are only −0.25 and −0.0009 respectively. Attitudes to friendliness are not closely related to respondents' length of residence in the area, as might be hypothesised. In this case the independent variable coefficient is −0.06. Also, views on the area's visual appearance were not influenced by gender as was supposed (female coefficient −0.23; male coefficient −0.08).

However, some significant results were obtained (Figure 5.4), the first major one being that residents do not consider the area aesthetically attractive. Only 13% of the responses indicated that the area is either moderately or very attractive. Conversely, a majority of the sample believed it to be in some degree unattractive, with a third having different attitudes. A chi-squared test (23.84 with four degrees of freedom) shows with 99.9% confidence that the population as a whole think that the local environment either is indifferent or in some degree unattractive. Negative attitudes are also evident with respect to the perceived dirtiness/cleanliness of the area. Probably a certain degree of colinearity cannot be excluded between this and the previous variable. However, the pattern of

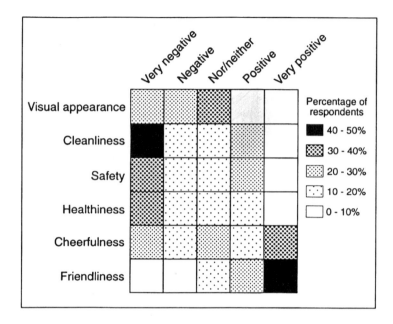

Figure 5.4. Perceived quality of the physical and human environment in Cattedown, Plymouth.

responses for each of them is different, the latter being more extremely felt, with over 40% of respondents considering their streets 'very dirty'. Attitudes to safety in the area produced a response pattern similar to the previous one. Almost 40% of respondents perceived the area as very dangerous, but 25% also felt it to be moderately safe. The chi-squared test shows this difference to be significant at the 99.9% confidence level. Conversely, however, chi-squared testing does not show that there are clear views about the area's healthiness and cheerfulness.

Clearer attitudes emerged once more when the analysis turned to whether the area is considered friendly. Nearly 50% of the respondents defined the area as 'very friendly' and another 25% described it as 'moderately friendly'. Therefore, nearly 75% had positive perceptions in this respect, and chi-squared tests confirmed that this pattern of responses is unlikely to have arisen by chance (47.13 with four degrees of freedom, significant at the 0.001 level). Summarising, there is much to suggest that the local inhabitants see the area as a problematic one. In fact, an overall description would be: an ugly, dirty and dangerous area, but with a good human environment.

DESIRABLE IMPROVEMENTS

As we have seen, 48% of the respondents claimed they were somehow irritated because of the economic activities carried out in the area. By calculating the standard error of the binomial frequency distribution, we may be 95.5% certain that this is true of between 59 and 37% of the full population. Therefore, at least a large minority of people and perhaps a clear majority perceive their local environment as damaged or at least diminished in its quality. Six main problems have been isolated, in order: noise; vibration; threat to safety by heavy lorries passing through the area; high speed of general traffic; pollution and dirt caused by local works; and intrusive parking by workers from outside the area (see Figure 5.6). Following an increase in these unpleasant conditions, according to some evidence, there has been a general withdrawal from the external space adjoining the houses: children are prevented from playing on the pavement, the residents no longer enjoy street conversation as they used to, because of traffic noise and a perceived lack of safety. In this respect, however, it is not possible to exclude recent changes in life-style.

To a great extent, as was expected, proposed improvements in the quality of the local environment stem from the perceived environmental problems. In order of frequency, the following have been proposed: some traffic-calming measures; relocation of the most incompatible activities, particularly those producing fumes and unpleasant smells; more frequent street sweeping; and general greening of the streets. A strong demand for improved policing has also been expressed, but this naturally lies well outside the responsibility of the planning system.

DERELICT PREMISES: BLEAK EYESORES AND TEMPTING OPPORTUNITIES

However strong feelings are on local environmental quality, everybody knows that the existing physical structure is not going to change very much. It is there to stay, at least in the medium term. Nevertheless, there are a number of premises in apparent dereliction in Cattedown that might potentially represent catalysts in land-use change. Question 4 identified perception of opportunities provided by derelict land, while question 5 identified localities considered appropriate for improvement. Together they give some insights into the perceived wishes of the community about how land uses should evolve, the degree of awareness about redundant space in the area and, indirectly, the level of attraction the waterfront exercises on local residents. Several major desired improvements in land-use patterns dominate local attitudes. The well-being of children is of wide concern, causing a high demand for safe recreational facilities. In addition, housing and leisure facilities are also considered very important. It is noteworthy that most respondents suggested that there is a need for cheap houses, but not council houses. This suggests the existence of a NIMBY (Not-In-My-Backyard) syndrome, perhaps connected with the bad name of some council estates. Generally, therefore, there is a desire to give the area an improved residential character.

Awareness of specific examples of derelict land appropriate for improvement appear dominated by the gas holder station between Clovelly Road and Shapters

Road, mentioned by 37% of respondents. The number of references to the waterfront or the now demolished power station that occupied a wide area along it was significant but unexpectedly low (25% of the total). Redundant space is extensive but was not picked out as an excellent opportunity. After decades without good access to the waterfront, it seems that local residents no longer have it on their mental maps. Physical topography is probably one of the causes: the Cattedown peninsula has two distinct zones. The one is mainly occupied by the residential built-up area that stretches to the ridge and looks northwards, towards the city. The other, that accommodates most of the industrial and commercial premises, slopes southwards, to the water. The general appearance of this industrial zone has a strongly repulsive atmosphere. But the main cause is probably the disappearance of the functional relationships between the community and the port, as is the case of most of the historic ports (Hoyle, 1988), to a degree that residents no longer frequent the industrial area or the waterfront itself.

PUBLIC PERCEPTIONS AND PLANNERS' PROPOSALS

The question to be considered now is the extent to which perceptions found in the survey are reflected in public plans for the area. Do local plans, developed on the basis of planners' external perception of the area, propose changes that will develop the area in ways which the local community feels desirable? Two recent plans have been considered. The first was produced in 1989 and proposed a joint venture between the public and private sectors, according to a pattern common to many waterfront redevelopment schemes. At an estimated cost of £60m, the city council and two developers, Bellway Urban Renewals and Carkeeks, proposed: the building of 500 new houses on the east side of Sutton Road; the refurbishing of 1000 dwellings; the development of a business centre in Coxside; a luxury hotel (140 beds) to be built at Queen Anne's Battery; construction of new road links; new industrial development sites made available through land reclamation from the Cattewater; a footbridge across Sutton Harbour and increased public access to the whole of the waterfront. The ambitions of this joint venture did not immediately and completely succeed, but some of the main points have since been included in the 1992 City of Plymouth Local Plan. This was published in March 1993 and, being the most recent plan, it is probably the best reflection of the planners' present perceptions of the area's needs. There is certainly significant overlap between local concerns and planning proposals. This is the case with respect to housing: in fact, the area east of Sutton Road will be reserved for residential development (Figure 5.5). In addition, proposals for new road links should, at least in theory, divert part of the traffic out from the residential area and channel it across the Cattedown industrial area.

Conversely, however, in a number of important ways the plan does not cover needs as they are locally perceived (Figure 5.6). First of all, as far as housing is concerned, it is evident that this project is unlikely to contribute significantly to

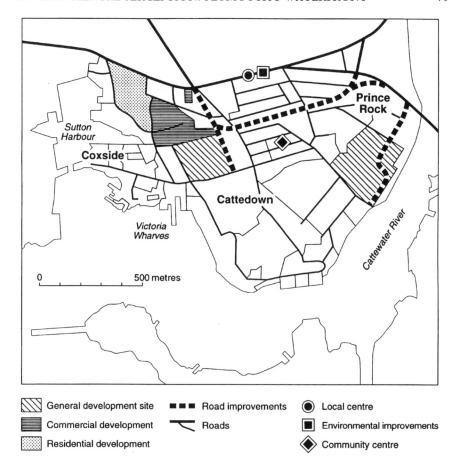

Figure 5.5. Planning proposals for Cattedown, Plymouth, 1993

the residential character of Cattedown itself. The new residential area will be physically isolated by economic land uses and will gravitate towards the tourist area of Barbican, alongside Sutton Harbour. Moreover, Cattedown is not considered by planners to be deficient in terms of recreational facilities for children. Most of the built-up area is within a 400 m radius from any such facility, the maximum recommended distance for this purpose. However, it is unlikely that providing facilities on the basis of a fixed Euclidean distance is going to be particularly successful in fulfilling perceived needs. In fact, despite a strong demand for recreational facilities for children, the few existing were constantly underutilised. This could lead such demand to be regarded as non-consistent, or alternatively the supply to be thought inappropriate. In fact, the demand is for close, safe places. In an area where safety is a major concern,

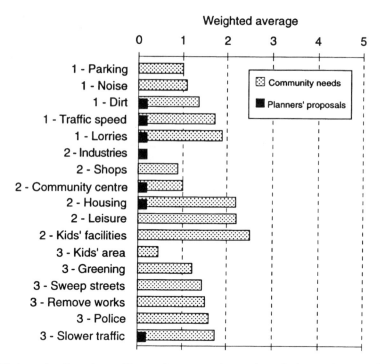

Figure 5.6. Cattedown, Plymouth: community needs and planners' proposals. Relative importance of (1) perceived environmental problems; (2) perceived opportunities for use of derelict land; (3) street improvements considered necessary.

evaluation of the level of provision of such facilities would probably be better carried out on the basis of some sort of cognitive distance (Gatrell, 1983).

In the local plan there is also no emphasis on using redundant space for general leisure and recreation. Derelict premises will be entirely devoted to industrial development. Additional industry was not something local people considered important. Visual appearance and sanitary conditions were also perceived as serious problems but were not addressed by planners' proposals. However, it must be recognised that some refurbishing of domestic façades was going on at the time of the survey, apparently with the help of the city council, in the form of low-interest loans. The major recreational project the plan proposes is a National Aquarium, but this is unlikely to meet the local demand for day-to-day leisure. An attempt to provide these opportunities is made by proposing a community centre, but the survey found little demand for this type of facility.

CONCLUSIONS

Overall, therefore, it must be concluded that a significant gap has been found between the aims and wishes of the local population and planners' perception of the area's needs. Even though a genuine willingness of the Plymouth Planning Department to involve local people in the planning process does exist, the investigation in Cattedown shows that there is still some room for improvement. A certain degree of overlap between the environmental perceptions and needs of the residents on the one hand and the planners' proposals on the other hand does exist, but the match is far from close. This failure might be partly due to objective constraints such as a shortage of financial resources or a lack of available space or of control of it. However, there are good grounds for believing that another important reason is a deficiency of detailed knowledge about local perceptions. Probably there is a weakness in the communication process between the public and the Planning Department. At the moment the process is based on sporadic information meetings and individual written complaints when new plans and proposals are published, but it is questionable whether these really provide individuals and communities with a full opportunity of expressing their views. It can be argued that many members of the public have in practice no access to the process. In fact public hearings, structured workshops, legal notices, information meetings, citizens advisory committees and written comments are all techniques that acquire opinions strongly biased in favour of the highly literate, highly verbal and highly informed and involved public. Also in this respect planning departments have, and legally must have, a passive attitude of just waiting for possible complaints by the public. We would be very surprised if private companies acted similarly. Instead they actively pursue research concerning the perceptions and needs of their markets. We usually would not consider this as a demonstration of social justice, but rather a way of pursuing a company's interest instead. Planning departments sometimes invest considerable capital with a high risk of missing their target. This chapter has tried to show that feasible alternative methods of collecting detailed information about perceptions and local needs do exist. Polls have a high degree of representativity, being able to reach all the different segments of the public.

In the past, planners have often evaluated urban environments merely on measures of physical characteristics. Dwelling densities, indices of overcrowding, lack of internal amenities, etc., are seen as objective variables of the quality of the environment. However, users often have different views on what really matters in the environment. As plant localisation, design and market research are part of complex strategies in the private sector, there is no reason why the public sector should not use similar strategies in its planning process. Of course, we have to recognise that very often planners do not have social research skills in their background, but this is not a good reason not to hire some skilled personnel in the planning teams. This is important on the grounds of both social justice and economic efficiency, especially nowadays when financial resources available to city councils are diminishing.

This leads us to a more general conclusion. It is well known that many waterfronts are undergoing substantial change. This opens new perspectives for the revitalisation of the urban structure of many cities. It is necessary to recognise, however, that the attitudes and needs of communities gravitating towards coastal zones are changing as well, and maybe even more quickly. The higher the well-being of a community, its level of education and the spare time available, the lower is the tolerance to developments that could threaten its quality of life. It is therefore necessary that the legislative structures recognise changes that are taking place and create flexible tools that might possibly forecast the emergence of conflicts by acquiring new conceptual instruments of investigation in a multidisciplinary and integrated approach to territorial planning.

REFERENCES

City of Plymouth (1989), *West Coast watershed: planning for Plymouth in the 1990s* (Plymouth: Plymouth City Planning Department and Royal Town Planning Institute).

City of Plymouth (1990), *1990 – Industry survey* (Plymouth: Plymouth City Planning Department).

City of Plymouth (1992), *Results from the 1991 Census of Population* (Plymouth: Plymouth City Planning Department).

City of Plymouth (1993), *City of Plymouth – Local Plan 1992: first alteration* (Plymouth: Plymouth City Planning Department).

Clout, H., Blacksell, M., King, R. and Pinder, D. (1985), *Western Europe: geographical perspectives* (London: Longman).

Gatrell, A.C. (1983), *Distance and space: a geographical perspective* (Oxford: Clarendon Press).

Hilling, D. (1988), 'Socio-economic change in the maritime quarters: the demise of sailortown', in Hoyle, B.S., Pinder, D.A. and Husain, M.S. (eds), *Revitalising the waterfront: international dimensions of dockland redevelopment* (London: Belhaven), 20–37.

Hoyle, B.S. (1988), 'Development dynamics at the port–city interface', in Hoyle, B.S., Pinder, D.A. and Husain, M.S. (eds), *Revitalising the waterfront: international dimensions of dockland redevelopment* (London: Belhaven), 3–19.

Lucia, M.G. (1994), 'Waterfront: una nuova frontiera per le città d'acqua', in Hoyle, B.S., Pinder, D.A. and Husain, M.S. (eds), *Aree portuali e trasformazioni urbane: le dimensioni internazionali della ristrutturazione del waterfront* (Milan: Mursia), 11–18.

Maguire, D.J. *et al.* (1987), *Plymouth in maps: a social and economic atlas* (Plymouth: Department of Geographical Sciences, Plymouth Polytechnic).

Pinder, D.A. (1981), 'Community attitude as a limiting factor in port growth: the case of Rotterdam', in Hoyle, B.S. and Pinder, D.A. (eds), *Cityport industrialisation and regional development* (Oxford: Pergamon), 181–99.

Pinder, D.A. (1984), 'Industrialisation and the quest for upward economic transition: an examination of development strategies for the Dutch Delta', in Hoyle, B.S. and Hilling, D. (eds), *Seaport systems and spatial change* (Chichester: Wiley), 277–301.

Pocock, D. and Hudson, R. (1978), *Images of the urban environment* (London: Macmillan).

Royal Society Study Group (1992), *Risk: analysis, perception and management* (London: The Royal Society).

6 Oil Industry Restructuring and its Environmental Consequences in the Coastal Zone

SARAH HARCOMBE and DAVID PINDER

Department of Geographical Sciences, University of Plymouth, UK

The major phase of seaport expansion which occurred in the post-war period up to the mid-1970s brought about substantial environmental losses in coastal zones. Throughout the developed world, thousands of hectares of land were consumed by port growth, the areas appropriated often being ecologically rich wetlands including saltmarshes, reedbeds and tidal mudflats. In many instances the latter provided extensive feeding grounds for wading and migratory birds (Pinder and Witherick, 1990). Because this was the era of the MIDA (Maritime Industrial Development Area), the new port areas that were created were usually colonised by major industries with their attendant pollution threats (Verlaque, 1981; Vigarié, 1981). Allied to these developments was the risk of environmental damage through cargo transfer spillages and other accidents as increasing numbers of ships, often carrying dangerous and toxic liquids, moved in and out of ports (Dudley, 1976). If governments, port authorities, shipping interests and industrial organisations recognised these problems at all, they were normally seen as the price to be paid for economic gains.

Since the mid-1970s circumstances have changed, and it is now arguable that these transport-related environmental pressures on the coastal zone have been ameliorated, at least in part. This reflects the fact that demand for seaport growth has abated in many parts of the world. It has also come about because of the introduction of stronger environmental regulations, some devised as a direct response to growing realisation of the impact of many seaport developments. This is well demonstrated by the introduction of very stringent controls in Rotterdam, where port-related environmental conflict first reached major proportions in the 1970s (Pinder, 1981). But in addition – and most importantly from the viewpoint of this paper – reduced pressures are a consequence of industrial restructuring in response to new demand conditions. Demand for continued industrialisation has fallen, and in some instances seaport industries have shrunk rapidly as disinvestment has become the order of the day. This

Cityports, Coastal Zones and Regional Change. Edited by Brian Hoyle.
© 1996 John Wiley & Sons Ltd.

deindustrialisation has commonly been economically disadvantageous for the localities in question because of, for example, employment losses and reduced port dues. But it is also arguable that this form of restructuring has in other respects been advantageous in that it has brought environmental gains. Industrial shutdowns may well be expected, for example, to lessen traffic congestion and reduce pollutant discharges, and this research has set out to test this proposition. To what extent can it be concluded that coastal zone environmental pressures, originally generated by port-related development, have been relieved by the restructuring process?

This question has been addressed by focusing on the oil refining industry which, in Western Europe at least, has been prominent in coastal zone deindustrialisation (Pinder and Husain, 1988; Pinder and Simmonds, 1994). As a response to a severe structural demand crisis in the early 1980s, more than one in five European coastal refineries has closed, primarily through the restructuring activities of the leading international oil companies (Figure 6.1). Within Western Europe the investigation reported here has concentrated specifically on the UK, where eight coastal refineries have closed since 1976. This is a third of all European coastal refinery closures and, as is indicated below, the varied nature of the sites in question provides ample opportunities to explore the implications of closure in a wide range of contrasting circumstances.

The argument offered is that environmental gains resulting from this form of port restructuring are indeed identifiable and, at least in the context of oil refining, are best considered in terms of the amelioration of permanent and intermittent environmental impacts. In addition, however, this argument is extended to propose that the environmental gain process is more complicated, and less guaranteed, than it appears at first sight. Above all this reflects two factors. One is that, although the term 'refinery closure' has an air of finality, it is by no means always the case that the cessation of normal refining means the end of all oil-related activities. As will be demonstrated, some activities and their associated environmental impacts may well continue. Second, while waterfront revitalisation is normally thought of as a process transforming old port areas close to the hearts of cityports (Falk, 1992; van der Knaap and Pinder, 1992), waterfront refinery sites are now beginning to prove attractive to a range of new activities. These typically bring with them different, yet still significant, environmental effects. Thus, the potential environmental gains to be made from refinery shutdowns in the coastal zone are often reduced because of the impact of surviving or replacement economic activity. In other words, gross gain becomes net gain. Before exploring this argument in full, however, it is necessary to provide an overview of the sites themselves and of the investigation on which the discussion is based.

ABANDONED COASTAL REFINERIES – A UK OVERVIEW

Details of the abandoned sites are provided by Table 6.1 and Figure 6.2. In terms

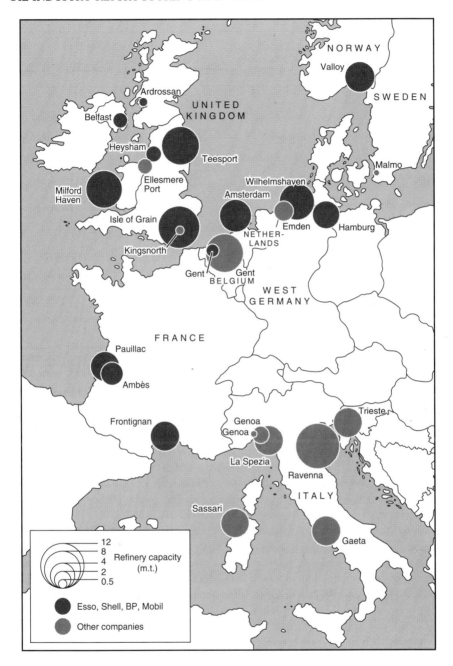

Figure 6.1. Coastal refinery closures in Western Europe, 1976–87

Table 6.1.   Characteristics of closed UK refinery sites

Location	Company	Refining capacity (m tonnes)	Area (ha)	Year closed	Nature of location
Isle of Grain	BP	11.1	600	1981	Rural
Milford Haven	Esso	8.6	500	1983	Rural, scenic (National Park)
Teesport	Shell	5.1	250	1985	Industrial seaport
Heysham	Shell	1.9	100	1976	Semi-urban
Belfast	BP	1.6	50	1982	Industrial port area
Ellesmere Port	Burmah Oil	1.3	100	1981	Industrial area
Kingsnorth	Berry Wiggin	0.3	40	1976	Rural, semi-industrial
Ardrossan	Shell	0.3	13	1986	Small town port area

Sources: Company data.

of pre-closure output, two main groups may be identified. Ardrossan, Ellesmere Port, Heysham and Kingsnorth were small enterprises refining less than two million tonnes of oil a year, and in some instances substantially less. In many ways these can be considered typical European closures, in that refining companies normally concentrated shutdown programmes on small and therefore inefficient undertakings. In contrast, the end of refining by Shell at Teesport, Esso at Milford Haven and BP on the Isle of Grain affected much larger activities, with BP's Isle of Grain shutdown being virtually the largest in Europe. Above all, these major closures reflected the leading refiners' willingness to view their production systems as a whole, and close even large advanced plants if they inhibited the optimisation of company output (Pinder and Husain, 1987).

Given this contrast in former production capacity, the abandoned sites resulting from closure naturally vary in scale. What is clear, however, is that almost all shutdowns have made available significant areas in the coastal zone. Five of the eight sites extend over 100 hectares (ha) or more, with the Esso Milford Haven and BP Isle of Grain sites accounting for no less than $5\,km^2$ and $6\,km^2$ respectively. Moreover, sites of 40 and 50 hectares at Kingsnorth and Belfast are not negligible. Linked with this question of scale is one of topography: the needs of refinery processing plants, and storage tanks for crude oil and refined products, normally required a site to have extensive level areas. Thus, refinery closures have released substantial sites that are potentially highly

Figure 6.2.   (*opposite*) Locational characteristics of closed UK refinery sites

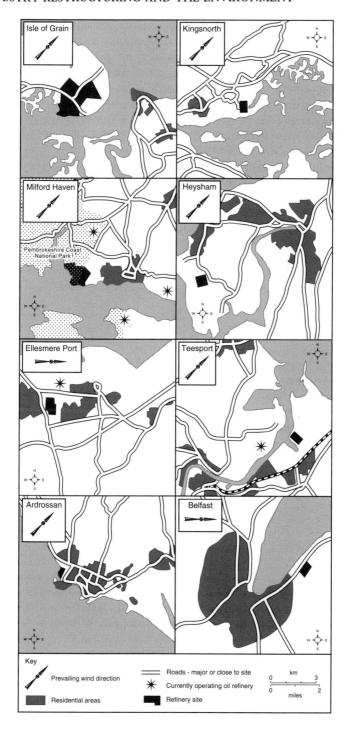

attractive for large-scale industrial or commercial activities. As Table 6.2 demonstrates, this has already stimulated significant re-use at several locations. This re-use is, of course, potentially associated with new environmental impacts.

A further relevant point is that several sites have what are essentially urban or semi-urban locations. This is particularly true of Ardrossan (where the 13-hectare site in the original port area lies close to the heart of this small town), but it also applies to Belfast, Teesport, Ellesmere Port and Heysham. The chief significance of this is that, when operating, the majority of the refineries were close to substantial populations. Consequently, it can be hypothesised that some environmental benefits of closure, such as the cessation of air pollution, should have been significant for local communities. Moreover, because none of the refineries were in basically unpopulated areas, this hypothesis can be extended to the remaining sites. Even Esso Milford Haven, the most rural of all the sites, lay only a few kilometres away from substantial post-war housing areas (Figure 6.2).

Quite apart from air pollution, the site with apparently the most obvious potential for reductions in pressure on the coastal zone environment was Esso Milford Haven (Figure 6.3). In this magnificent coastal setting, vessels carrying crude oil and oil products posed a frequent threat from spillage; cooling and process waters from the refinery flowed back into the Haven; and the site intruded significantly into the landscape of the Pembrokeshire Coast National Park. (Remarkably, the boundary of this Park passed through the refinery itself, and through the nearby Amoco plant; Figure 6.4.) However, Esso Milford

Table 6.2. Economic activity on former refinery sites

	Surviving from refining era	New	Approved
Isle of Grain	Oil products storage	Container/coal terminal CCGT[1] power station	LNG[3] import
Milford Haven			
Teesport		Car export terminal (Nissan)[2]	
Heysham		Industrial estates Solvent recovery	
Belfast	Oil products storage		
Ellesmere Port	Lubricant production		
Kingsnorth	Bitumen production	Industrial estate	
Ardrossan			

Sources: Company data, local authorities and field survey.

[1] Combined cycle gas turbine.
[2] Subsequently transferred to a competing port, April 1994.
[3] Liquefied natural gas.

Figure 6.3.   Esso's Milford Haven refinery in its coastal setting

Haven's obviously sensitive site should not be allowed to obscure other potential gains for the natural environment elsewhere. All sites were in fact at risk from tanker spillages, and all sites had the potential to pollute local waters through effluent emissions. In addition, although they lacked the obviously sensitive setting of Esso Milford Haven, some refineries had in their immediate vicinity what were undoubtedly delicate ecosystems. This was particularly true of the two locations on the Isle of Grain, BP Grain and Kingsnorth. Saltmarshes in their vicinity included a major Site of Special Scientific Interest (SSSI); provided an important feeding ground for wading and migratory birds; and belonged to the country's dwindling stock of wetland.

Research design

Research into the environmental consequences of these shutdowns adopted three approaches. First, surveys were conducted among a wide range of organisations expected to hold information on the pollution impact of the eight refineries and on improvements achieved since closure. Participants in these surveys included oil companies, port authorities, environmental health departments, planning authorities and pollution control agencies such as the National Rivers Authority and Her Majesty's Inspectorate of Pollution. Altogether, 40 respondents were contacted in this aspect of the research. Second, a literature review was under-

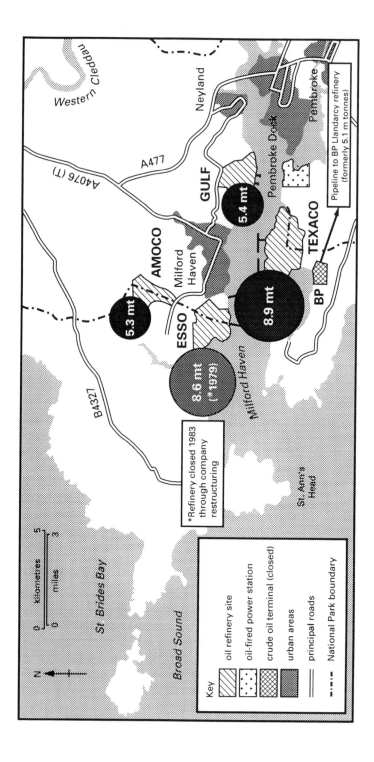

Figure 6.4. Oil industry development around Milford Haven

taken to identify studies relating either to the environmental impacts of the eight refineries in question, or to the general environmental consequences of oil refining and refinery closures. Third, to assess the environmental implications of post-closure revitalisation, surveys were conducted to identify and evaluate all economic activities operating on the former refinery sites in mid-1993. These surveys brought together publicly available documentation (such as planning assessments and Environmental Impact Statements); the views of key officials (particularly in the planning system); attitudinal investigations among local communities; and information supplied by the replacement economic activities themselves.

Despite the breadth of the investigation, the available information was found to be fragmentary. In part this reflected the fact that, although relevant data had in some instances been collected while refineries were operational, they were destroyed after closure. For example, the last UK refinery to shut was Ardrossan in 1986, and files relating to air pollution emissions from this plant were destroyed two years later. In addition, however, the fragmentary nature of the data reflects a relative absence of previous academic interest in the subject, as well as the problem that relatively little environmental monitoring has traditionally been carried out in the UK. This latter point highlights one of the recurrent problems of coastal zone management, namely that planning is frequently hampered by incomplete and poorly coordinated information sets (Gubbay, 1990; House of Commons Environment Committee, 1992). Despite these difficulties, however, it is possible to identify with reasonable clarity a range of preliminary findings relating to the environmental outcomes of the refinery closure movement. This is particularly true if the findings are considered in terms of persistent and intermittent pollution, and with respect to environmental gain and loss.

CLOSURES AND ENVIRONMENTAL GAIN

Persistent pollution trends

Operational refineries are commonly expected to be sources of everyday air pollution. Particulates and combustion products such as sulphur, nitric oxides and carbon oxides all have the potential to impose persistent local impact in the coastal zone. So, too, do hydrocarbon emissions from storage tanks. In practice, however, there is little evidence that closures were in fact associated with major local gains in this respect. For example, none of the Environmental Health Officers contacted during the survey raised persistent air pollution as an issue linked with the former operations of any of the eight refineries. When interpreting this finding, some caution is of course necessary. Local communities – perhaps encouraged by a measure of economic dependence on the refineries – may have adjusted to their activities so that 'background' pollution was ignored. At some urban–industrial sites, such as Teesport and Ellesmere Port, specific refinery pollution may not have become an issue because it merged with emissions from other sources. And at four refineries – Esso Milford Haven, Teesport, Belfast and Ellesmere Port – high stacks deliberately designed to disperse combustion products may have kept local pollution under control at the expense of distant areas. None the less, it is striking that years of refining activity failed to create obviously hostile attitudes among communities. This in turn suggests that – although air pollution may have been a reality – persistent improvements in atmospheric quality that were perceptible to local people after the closures were marginal rather than fundamental.

Similarly, there is little to indicate that the persistent severe pollution of coastal waters by process effluents has been a general problem giving scope for large post-closure environmental gains. This is not to argue that process discharges were of no significance: there is certainly evidence that marine environments in close proximity to refinery discharge points are adversely affected by effluents. Indeed, at two of the refineries in question (BP Isle of Grain and Esso Milford Haven) detailed investigations while the installations were still in use uncovered ecological abnormalities around outfall pipes (Wharfe, 1975; Dicks and Levell, 1989). Yet these results applied at the very local scale, to the extent that adverse effects could be identified within a 100-metre radius of the discharge points. This restricted spatial impact is also underlined by ecological surveys conducted by Dicks and Levell (1989) around the Esso Milford Haven plant before and after its closure (Figure 6.5). Following the shutdown an increasing limpet population was identified at a monitoring station just 30 metres from the outfall. But no such discernible trend was apparent at a second monitoring station which, although more distant, was still only 225 metres from the discharge point. What may be added is that any effects on fauna often appear to be the result of localised changes in salinity, rather than of contaminants in the outflow. To quote Petripoon and Dicks (1982), 'The effects of ... effluent discharge cannot be considered as significant damage, although they are biologically interesting'. It follows, therefore, that gain from the cessation of effluents must also be limited.

Conversely, there are considerably stronger indications that local environments are able to benefit from the amelioration of two further types of environmental impact – visual intrusion and road traffic pressures. Refineries are often criticised for their visual intrusion in the coastal landscape, and it might be expected that a long-lasting consequence of closures would be further environmental loss resulting from the dereliction of installations. This would be a particularly important consideration at all refineries outside urban areas, but particularly at Esso Milford Haven. However, the industry has shown itself to be highly effective at site clearance. Some processing plant has been sold and dismantled for use elsewhere, while much of the remainder has simply been disposed of for its scrap value. Similarly, surplus storage tanks have been demolished. Curtailed visual pollution has therefore been a typical form of environmental gain which, it may be noted, has also brought other potential

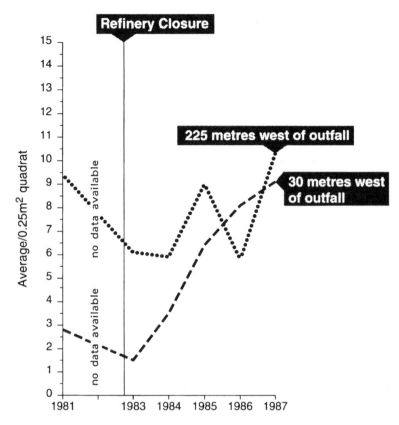

Figure 6.5. Limpet population changes around the Esso Milford Haven outfall. (after Dicks and Levell, 1989)

benefits. These chiefly relate to the control of post-closure pollution hazards associated with oil residues in refinery processing plant. Removal of the plant and redundant storage facilities eliminates the long-term possibility of pollution caused by residue seepage into the environment.

In the case of road traffic pressures, the significant gains are linked with journeys to work and product distribution. Journey-to-work data produced by the survey suggest that no more than a third of a refinery's labour force is likely to use public transport and, depending on the plant's location, the proportion may be much less. This is particularly the case when sites lie well outside port and urban areas. Thus, linked with closures, there may be significant gains to be made in terms of reduced car usage, although the extent to which this is the case naturally varies according to circumstances in specific refineries. In some the scope for gains is limited, partly because of their restricted size, and also because past economic and safety pressures encouraged the minimisation of labour forces. For example, five of the eight closures employed less than 300 workers at the time of the shutdown, and two had less than 100. But the Shell plant at Teesport employed almost 600, while for Ellesmere Port and BP Isle of Grain the figures were 1000 and 1300 respectively. As Table 6.3 indicates, in most scenarios employment losses of between 600 and 1300 can be expected to reduce peak-time traffic flows around sites by at least several hundred vehicles, even allowing for substantial use of public transport and car sharing.

So far as product distribution is concerned, much again depends on the organisational details of individual refineries. The closure of Esso Milford Haven, for example, offered virtually no benefits in terms of reduced goods traffic on the road system. This was because 60% of the products were marketed via a long-distance pipeline (which linked the refinery to Birmingham, Nottingham and Manchester), and partly because 30 to 35% of output was moved by sea. Small amounts also left the area by rail, giving very little scope for distribution by road tanker. Esso Milford Haven was, however, exceptional in this respect, and the large majority of refineries have relied much more heavily on road transport. Moreover, in some instances this tendency has been reinforced by company decisions to use refineries as import points for products refined elsewhere. In these circumstances, imports have normally been marketed regionally by road, together with the local refinery's own output. A good example of this is provided by Ardrossan, which Shell used as its distribution centre for South West Scotland. The Ardrossan refinery itself was small and highly specialised, importing no more than 250 000 tonnes of heavy crude oil each year for use in bitumen production. But finished products from other Shell refineries were brought in by sea and distributed throughout the region by 38-tonne road tankers. In total, these covered more than 1 million miles (1.6 million km) per year. Local environmental gains here have certainly been

Table 6.3.   Alternative scenarios for car-based commuting

Scenario	Car usage	
	600-employee refinery	1300-employee refinery
No use of public transport; no car sharing	600	1300
No use of public transport; car sharing by 50% at average 2.5/car	420	910
Public transport carrying 33.3%; no car sharing	400	870
Public transport carrying 33.3%; car sharing by 50% of remainder at average 2.5/car	280	610

Source: Authors' calculations.

considerable, not least because the location of the refinery – in the old harbour area – unavoidably routed road-tanker traffic through and past heavily populated parts of the town (Figure 6.2).

Intermittent pollution

Although very limited gains appear to have been made from the eradication of *persistent* air and water pollution, it is arguable that the elimination of largely *random* water and air pollution events has brought appreciable benefits in localities affected by closures.

Intermittent water pollution has three main causes: spillages during transhipment operations at a refinery's jetty; collisions in harbour areas or adjacent inshore waters; and deliberate discharges by oil or oil products tankers during, for example, illegal tank flushing operations (Dudley, 1976). Comprehensive data relating to these sources are unavailable, but one study highly relevant to this paper is that by Hobbs and Morgan (1994). This covers accidental discharges into Milford Haven, a port area long known for its attempts at good management in this sphere. In some respects the data presented by Hobbs and Morgan suggest that the gains to be derived by eliminating spillages are limited. For example, although the average number of spillages between 1961 and 1990 was 39 per year, in most years less than 50 tonnes entered the Haven for this reason (Figure 6.6B). Similarly, the number of spills per 100 ships handled fell significantly, reaching only 0.8 in the period 1986–90, while the number of spills per million tonnes of oil handled fell by two-thirds. In other respects, however, the data also show that in the long run even an extremely well-managed port may be unable to avoid significant spillage problems. In this connection the most outstanding point is that, although what are classified as major incidents have recently involved smaller spillages than in the 1960s and 1970s, they have become no less frequent (Figure 6.6A). Oil refinery closures are therefore likely to produce environmental gains through the elimination of this hazard, as well as through a reduction in minor incidents.

Much of the ecological evidence concerning the gains to be made from the cessation of spillages also comes from Milford Haven (Baker, 1976; Dicks, 1976; Petpiroon and Dicks, 1982; Howard et al., 1989). This indicates that long-term damage can be done to flora such as seagrasses, the density of which can be significantly lowered, leading to denudation during subsequent winter feeding by wildfowl. When this occurs the result may be bare sediment beds with a vegetation recovery time of up to 100 years. Such damage to flora is likely to be greatest when spillages involve refined products, which tend to be more toxic than crude oil. Similarly, the use of dispersants is also far more damaging than mechanical means of recovering floating pollutants. But what is also shown by research in Milford Haven is that the effects of spillages here may be limited because of the locality's particular conditions (Dicks, 1976). The general absence of a muddy shoreline in the Haven, the high rate of flushing and an alert port

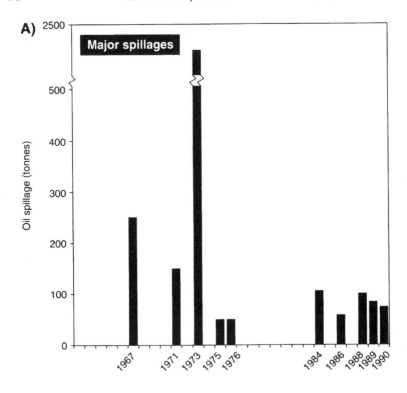

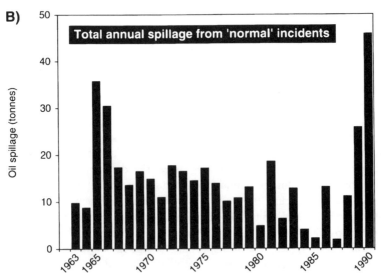

Figure 6.6. Crude oil and petroleum spillages in Milford Haven: 'normal' levels and major incidents, 1963–90. (after Hobbs and Morgan, 1994)

management all serve to limit the impact of incidents. The implication of this is that the environmental gains to be made from the cessation of spillages are likely to have been significantly higher at other sites, such as BP Isle of Grain and the nearby Kingsnorth site. This chiefly reflects the fact that the extensive saltmarshes in these localities, which in the prevailing low-energy environment are not subject to intensive flushing, are intrinsically more vulnerable than the 'hard' shore environment of the Haven.

The gains to be made from the elimination of intermittent air pollution are less clear-cut. All refineries are from time to time affected by periods of high atmospheric pressure, and also by temperature inversions. In both sets of circumstances, normal wind patterns – which dilute and disperse emissions – are overcome and replaced by still-air conditions likely to trap air pollution in the vicinity. In extreme conditions the results have been known to be severe (Pinder, 1981, 192–3). Against this background, and given the fact that significant populations lived in the neighbourhood of all the closed refineries (Figure 6.2), it might be expected that closure would yield substantial gains. Yet references to this problem by Environmental Health Officers in the survey were few, and tended to be couched in rather low-key terms such as 'occasional odour complaints'. While this may accurately reflect the absence of a significant problem, it may again indicate that local communities had adapted to this form of pollution, accepting it as normal. Progress on this specific issue is unlikely to be made until investigations are conducted into community attitudes to pollution from refineries still in production.

GROSS OR NET ENVIRONMENTAL GAIN?

So far it has been assumed that all the potential environmental gains to be made from the cessation of refining are likely to be realised following a closure. From the survey, however, there is clear evidence that actual gains may fall significantly short of the theoretical maximum.

Continuing oil-related activity

One reason is that, while mainstream oil refining may cease, a company may also retain a minor part of its production activity. This can occur in response to local market conditions, as at the Kingsnorth refinery, where bitumen production has been retained (Table 6.2). This site is, of course, well placed to exploit South East England's strong demand for bitumen in the road maintenance and construction industries. Limited continuing production may also ensure that a company maintains its output of special products, as with lubricant production at the Burmah Ellesmere Port site. Such a strategy may be of considerable importance for the refinery, as this example in fact shows: in recent years the profitability of high-value lubricants has been central to the company's success.

Quite apart from the maintenance of specialist production, companies may also continue to use part of a former refinery site as a regional distribution centre. This strategy acknowledges that a regional market continues to exist. Even though local production may be uneconomic, past company investments in infrastructure such as storage tanks and jetties provide a valuable base which the firm may use for distribution in order to protect or even enhance its regional market share. This is well illustrated by the post-closure history of BP's Belfast refinery, where the company has retained much of the original site for use as an import terminal for the whole of Northern Ireland. This now receives around 2.2 million tonnes of oil products annually, compared with approximately 1 million tonnes of crude oil that were imported by the refinery in 1980, shortly before it closed.

Whatever a company's motivation for maintaining activity at a site, there are implications which reduce the environmental gains that would be attainable through total closure. Specialist processing may be linked with continuing air pollution, even though the problem is likely to be limited in scale. Retained storage and processing facilities are visually intrusive. And, most importantly of all, specialist production and the retention of sites as regional distribution centres have substantial implications for transport-related environmental problems. These partly relate to marine pollution: product distribution depots normally generate frequent coastal tanker movements, maintaining the probability of accidental spillages. But there are also major consequences in terms of land-based pollution. Virtually all bitumen, for example, is marketed by road, as are most refined products passing through regional distribution centres. Thus, returning to the example of BP Belfast, this site is now the focus for 13 000 return road-tanker trips each year.

Alternative activities and environmental impact

Continued use by aspects of the oil industry is not the sole factor likely to reduce significantly the environmental gains to be obtained from the cessation of mainstream refining. Instead, these gains are also likely to be eroded as new economic activities colonise the abandoned sites, most of which are likely to be zoned for industry or distribution as a result of their industrial past. To a degree, the impact of new activities may be to accentuate air or water pollution risks. Part of Shell's former Heysham site, for example, has attracted a solvent recovery plant. Although this is well suited to the infrastructure left by the defunct refinery, the firm itself indicated that the choice of this location also reflected concern that this was a potentially polluting development which would encounter considerable opposition if planning applications were made for other sites in the Heysham area.

Above all, however, the most consistent environmental loss likely to arise from an influx of new activities is – once again – the impact of road traffic. One major reason is that many alternative land uses are considerably more labour intensive than oil refining, the history of which has been inextricably connected with the substitution of capital for labour. Thus, apparently innocuous developments such as light industry may greatly accentuate journey-to-work movements on the local road system. The extent to which this may occur is demonstrated by a new industrial estate on the small Kingsnorth site. Although as yet this occupies only half the available land, its labour force (at least 150 people) is three times the number formerly employed by the refinery. Similarly, three industrial estates occupying parts of the larger Heysham site employ almost 600 workers, compared with the earlier refinery's labour force of only 200.

In addition, environmental loss reflects the fact that abandoned refinery sites – offering extensive stretches of prepared level land – are starting to prove attractive to land uses generating substantial heavy goods vehicle (HGV) traffic. The impetus that can be given to HGV movements is well demonstrated by developments on the Isle of Grain site (Figure 6.7). Current work on a new gas-fired power station is naturally the source of large-scale construction traffic, but completed projects at Grain also underline the capacity of the revitalisation process to generate major permanent traffic flows. The outstanding example of this is Grain's Thamesport development, a project combining advanced container handling facilities and a major coal import terminal. Although containers and coal will naturally arrive at the site by sea, and some domestic movement may involve sea or rail transport, in the early years of its operation Thamesport's

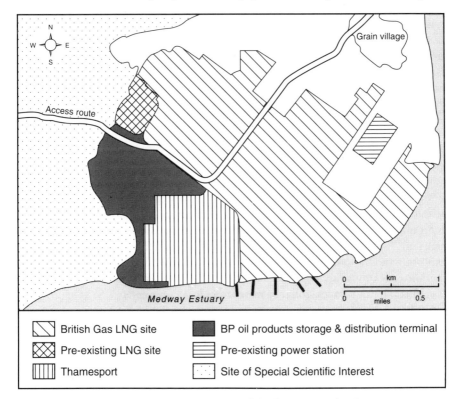

Figure 6.7. Economic revitalisation of the former BP Isle of Grain site

demands on the local road system are already proving obtrusive. Kent County Council has predicted that in 1996 Thamesport will produce 950 HGV movements, and at least 1100 light vehicle movements, per day. In the long term at least half the imported coal (a minimum of 250 000 tonnes per year) is likely to move by road, and Thamesport's total impact on the road network will far exceed that of the original refinery. Clear evidence that heavy goods traffic is already producing undesirable consequences is provided by community-based investigations conducted in this area as part of the current research. In a survey of 139 households in Grain village, 95% of respondents believed new developments at the refinery site to be the cause of excessive road traffic, and three-quarters considered this to be the result of growing HGV activity. In contrast, only 10% who had lived in the area when the refinery was operational described traffic at that time as very heavy. Even allowing for retrospective use of rose-tinted spectacles by many respondents, this discrepancy strongly suggests that sharply increased pressure on the road system has occurred.

CONCLUSIONS

It is evident that the relationships between coastal zone deindustrialisation and environmental gain are less clear-cut than they appear at first sight. What is also clear is that attempts to explore these relationships are, in the UK at least, currently hampered by data availability. Much of the information ideally required to elucidate matters is not gathered systematically. Moreover, at least in some instances, potentially valuable data have been destroyed because the relevant authorities envisaged no further use for them. Despite these obstacles, however, a number of conclusions can be proposed.

Although intuitively it might be expected that the primary gains to arise from closures would be the decline of everyday air and water pollution, in reality the opportunities for these types of improvement are limited. Refineries are certainly not pollution free, but it would appear that emission controls have been strong enough to prevent consistent degradation of air or water quality. Instead, the indications are that the total closure of a refinery is likely to benefit local environments and communities in four other respects. One is reduced landscape intrusion. A second is the avoidance of intermittent water pollution, normally the result of handling mishaps at or near the jetty. A third may be the abatement of intermittent atmospheric pollution, in most instances previously caused by unpredictable and uncontrollable meteorological conditions. The fourth – and certainly the most consistent – is the reduction of road traffic and its associated environmental disadvantages. This relief reflects both the decline of journey-to-work movements (dominated by the private car) and the end of product distribution by road tanker.

However, because refining companies do not always abandon redundant sites completely, the actual environmental gains may be significantly less than the potential improvements. Continued use may amount to limited specialist processing, with the attendant possibility of air or water pollution. But the reduction of potential environmental gain is likely to be seen most clearly through the retention of sites as regional distribution centres for refined products brought in by sea. On the one hand, this maintains the risk of occasional water pollution through spillages, especially as the number of incoming products tankers is likely to be greater than when imports are confined to crude oil. On the other, it ensures that, on a daily basis, local road systems and nearby communities continue to be exposed to the consequences of frequent road-tanker movements.

Additionally – even if all potential gains arising from refinery closures were to be achieved – this progress would in the long run be likely to be eroded by the movement of new activities to the redundant sites. Such revitalisation is, of course, very much in the interests of site owners wishing to capitalise on unused assets. It is also to the advantage of local authorities concerned to create employment. Once again, the primary environmental impacts imposed by new activities are likely to be transport-related nuisances and pollution. Replacement journey-to-work traffic may exceed that generated by the former refinery, especially if re-use results in relatively labour-intensive light industrial development. Perhaps most seriously of all, however, large-scale activities well suited to the revitalisation of refinery sites (such as container transhipment, bulk terminals and the car import and export trade) may emerge as sources of intense HGV traffic flows.

This chapter has, it is true, been based on the study of a single industry for which only partial information is available. Further research into the consequences of the upheavals in oil refining is desirable, as is the extension of investigations to other industries that have undergone or are experiencing restructuring. Until this work is complete, the findings presented here must be considered provisional. Despite these limitations, however, there are clear indications that for coastal zone management the outcomes of industrial restructuring may be significantly more complex than appears at first sight. The concept of environmental gain associated with deindustrialisation is certainly attractive, but must be treated with caution. In many instances the gross potential gains are in reality likely to become net gains because of continuing site usage; and, particularly if a site is located in an economically active region, there is the danger that gains made through industrial shutdown will be offset by losses as new activities flow in. The economic forces involved in both decline and revitalisation are powerful, and their successful harmonisation with coastal zone environmental protection provides a major challenge that has yet to be overcome.

REFERENCES

Baker, J. M. (1976), 'Ecological changes in Milford Haven during its history as an oil port', in Baker, J. M. (ed.), *Marine ecology and oil pollution* (London: Applied Science Publishers), 55–66.

Dicks, B. (1976), 'The applicability of the Milford Haven experience for new oil terminals', in Baker, J. M. (ed.), *Marine ecology and oil pollution* (London: Applied Science Publishers), 67–79.

Dicks, B. and Levell, D. (1989), 'Refinery effluent discharges into Milford Haven and Southampton Water', in Dicks, B. (ed.), *Ecological impacts of the oil industry* (London: John Wiley and Sons for the Institute of Petroleum), 287–316.

Dudley, G. (1976), 'The incidence and treatment of oil pollution in oil ports', in Baker, J. M. (ed.), *Marine ecology and oil pollution* (London: Applied Science Publishers), 27–40.

Falk, N. (1992), 'Turning the tide: British experience in regenerating urban docklands', in Hoyle, B. S. and Pinder, D. A. (eds), *European port cities in transition* (London: Belhaven), 116–36.

Gubbay, S. (1990), *A future for the coast? Proposals for a UK coastal zone management plan* (Ross-on-Wye: Marine Conservation Society for the World Wildlife Fund).

Hobbs, G. and Morgan, C. I. (eds) (1994), *A review of the current state of environmental knowledge of the Milford Haven waterway* (Orielton: Field Studies Council for the Milford Haven Waterway Environmental Monitoring Steering Group).

House of Commons Environment Committee (1992), *Coastal zone protection and planning* (London: HMSO).

Howard, S., Baker, J. M. and Hiscock, K. (1989), 'The effects of oil and dispersants on seagrasses in Milford Haven', in Dicks, B. (ed.), *Ecological impacts of the oil industry* (London: John Wiley and Sons), 61–96.

Petripoon, S. and Dicks, B. (1982), 'Environmental effects (1969 to 1981) of a refinery effluent discharged into Littlewick Bay, Milford Haven', *Field Studies 5*, 623–41.

Pinder, D. A. (1981), 'Community attitude as a limiting factor in port growth: the case of Rotterdam', in Hoyle, B. S. and Pinder, D. A. (eds), *Cityport industrialization and regional development: spatial analysis and planning strategies* (Oxford: Pergamon), 181–99.

Pinder, D. A. and Husain, M. S. (1987), 'Innovation, adaptation and survival in the West European oil refining industry', in Chapman, K. and Humphrys, G. (eds), *Technical change and industrial policy* (Oxford: Blackwell), 100–120.

Pinder, D. A. and Husain, M. S. (1988), 'Deindustrialisation and forgotten fallow: lessons from Western European oil refining', in Hoyle, B. S., Pinder, D. A. and Husain, M. S. (eds), *Revitalising the waterfront: international dimensions of dockland redevelopment* (London: Belhaven), 232–46.

Pinder, D. A. and Simmonds, B. (1994), 'Crisis intensity, industrial restructuring and the transformation of West German oil refining', *Erdkunde 48*, 121–35.

Pinder, D. A. and Witherick, M. E. (1990), 'Port industrialisation, urbanisation and wetland decline', in Williams, M. (ed.), *Wetlands: a threatened landscape* (Oxford: Blackwell), 234–66.

van der Knaap, G. A. and Pinder, D. A. (1992), 'Revitalising the European waterfront: policy issues', in Hoyle, B. S. and Pinder, D. A. (eds), *European port cities in transition* (London: Belhaven), 155–75.

Verlaque, C. (1981), 'Patterns and levels of port industrialisation in the western Mediterranean', in Hoyle, B. S. and Pinder, D. A. (eds), *Cityport industrialization and regional development: spatial analysis and planning strategies* (Oxford: Pergamon), 69–85.

Vigarié, A. (1981), 'Maritime Industrial Development Areas: structural evolution and implications for regional development', in Hoyle, B. S. and Pinder, D. A. (eds), *Cityport industrialization and regional development: spatial analysis and planning strategies* (Oxford: Pergamon), 23–36.

Wharfe, J. R. (1975), 'A study of the intertidal macrofauna around the BP refinery (Kent)', *Environmental Pollution 9*, 1–12.

Part III

ECONOMIC DIVERSIFICATION
IN DEVELOPING AREAS

7 Balkan Transport and Cityport Development in an Era of Uncertainty

DEREK HALL

Department of Leisure and Tourism Management, The Scottish Agricultural College, UK

THE CENTRAL AND EAST EUROPEAN CONTEXT

Despite competition between the major Western economic powers to establish spheres of influence in the region, foreign direct investment (FDI) in Central and Eastern Europe (including the former Soviet Union) has been stagnating, by the end of 1994 amounting to $22.8 billion since the fall of communism (Keay, 1995). None the less, economic and political restructuring, German unification, and increasing integration and rationalisation in Western Europe have been exerting a number of major impacts on cityport and maritime transport development in the region (Hall, 1992, 1993b).

Expansion and reorientation of east–west trade has contributed to a hastening of the restructuring of European road and rail systems, with considerable investment being committed to upgrading networks to and within Central Europe. Development of bimodal transport systems such as road-railers has required infrastructural upgrading between ports and their landward hinterlands and within ports themselves, not least to meet inevitable container traffic growth. Previous capital shortages and bureaucratic inertia were exacerbated by a basic lack of container terminal facilities both at ports and within rail networks, high degrees of manual labour, container shortages, inadequate repair and manufacturing facilities, and limited logistics technology.

A reinvigorated cityport focus of economic activity is emerging. Substantial possibilities exist for the development of value-added processing for export in a number of ports. A number of free ports/free trade zones, offering a variety of tax concessions for investment, together with relaxed import procedures and lower customs duties, have been established to attract foreign capital investment, employment generation and skill enhancement. Such zones, however, are likely to become progressively less attractive to Western companies as they proliferate east of the Elbe.

Cityports, Coastal Zones and Regional Change. Edited by Brian Hoyle.
© 1996 John Wiley & Sons Ltd.

In the medium term at least, following overland infrastructural improvements between West and East Europe, cargoes moving to and from Central Europe and the former Soviet Union will have a tendency to use West European ports such as Hamburg or Trieste, thereby replicating pre-war spatial patterns.

In the earlier months of their post-communist transition, Central Europe's land-locked states saw the need to establish hub roles within the new international waterway system brought about by the Rhine–Main–Danube canal link in the North Sea–Black Sea waterway system. However, the fact that the Danube passes through Serbia (Figure 7.1) has considerably constrained long-distance waterway movement across Europe in the short- to medium-term, and has thus reduced any value the new canal link might have had.

New transport infrastructures and patterns of movement were required to circumvent the Yugoslav wars of succession and to comply with UN sanctions imposed against Serbia/Montenegro. Albania, the least developed country of Europe, found itself in the first half of the 1990s both emerging into the light of the 'new world disorder' from almost half a century of Stalinist rule, and becoming a beachhead for the development of such new Balkan transport links. This somewhat precarious position sets the theme for the focus of the rest of this chapter.

THE ADRIATIC GATEWAY TO THE BALKANS

Through the upland barriers of inland former Yugoslavia and Albania, only four natural major routes exist to link the Adriatic with the interior of south-eastern Europe. Three of these pass through Albania (Figure 7.2): (a) along the river Drin to Kosovo; (b) via the Shkumbin valley across central Albania to Macedonia and onwards to Greece and the port of Thessaloniki; and (c) along the Vjosë river system to Ioannina in Greece. The fourth route follows the Neretva river across Dalmatian Croatia into Bosnia, while a further man-made route was forged in the 1970s with the construction of the Belgrade–Bar railway, providing land-locked Serbia with an Adriatic outlet via Montenegro (Wilson, 1971; Singleton and Wilson, 1977).

Control of these routeways and their Adriatic ports has been crucial to, and symbolic of, the changing political geography of the region. Cvijić (1918) pointed to the differing forms and effectiveness of control exerted in Balkan history, distinguishing Western civilisations, such as those of Greece and Venice, which merely gained footholds along the Adriatic coast to stimulate trade. The Romans, by contrast, penetrated inland, establishing communication systems based on the four major routeways. Subsequently, the Byzantine and Ottoman 'oriental' civilisations' domination of the Balkan interior prevented easy penetration from the Adriatic by other powers. Finally, 'patriarchal' societies – Illyrian and Slav – employed Balkan topography to remain relatively isolated, often using potential through routes such as valley bottoms as territorial boundaries, thereby reducing their utility.

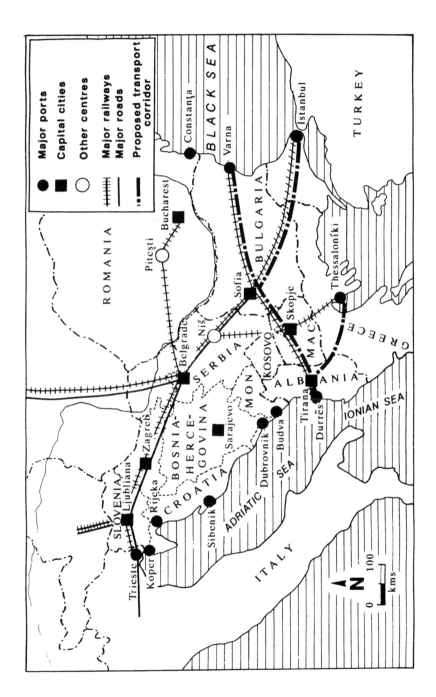

Figure 7.1. The Balkans of the mid-1990s

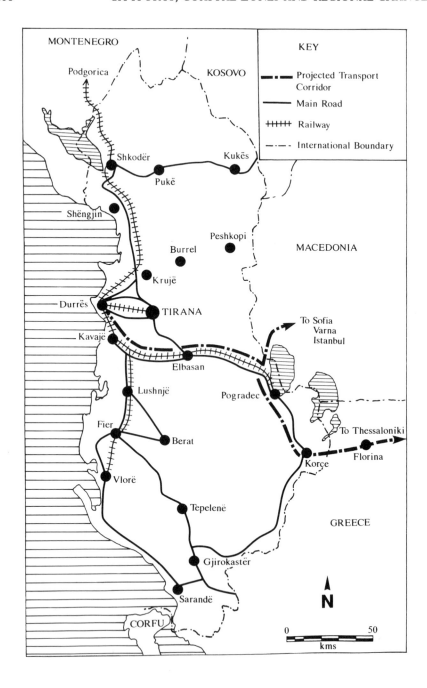

Figure 7.2. Albania: ports and major routeways

Thus the significance of routeways across Albanian lands reflected, on the one hand, the desire of Adriatic civilisations to gain access to Constantinople/ Byzantium/Istanbul and to the trade of the Balkan peninsula, and the response of Balkan powers to block such access and to prevent footholds along the coast. Such geostrategic considerations have extended into the present century.

Most important of the penetrating routes has been that following the Shkumbin river across central Albania. Along it, the Via Egnatia, built by the Romans 'to guarantee the security of Macedonia' (Hösch, 1972, 35), linked Rome to Byzantium, by way of Brindisi, Ohrid, Monastir (Bitolj), Thessaloníki and Kavalla. The site of Albania's present major port, Durrës (Dyrrachium/ Epidamnus), was an important transhipment point on this route for goods and people crossing the Adriatic (O'Sullivan, 1972; Raven, 1993).

TRANSPORT AND REGIONAL CHANGE: ALBANIA'S ELEVATED STATUS

Bottlenecks and constraints

In the 20th century, Durrës was handling half of Albania's imports in the 1930s, with 25 000 tonnes of cotton and woollen goods, timber, cement, petroleum products, iron and steel, and food coming into the port in 1937. Exports accounted for 32 500 tonnes of olives, hides, skins, eggs and 24 000 head of cattle. Infrastructural facilities were minimal: 2000 tonnes was the maximum which could be cleared in a day from the port landwards, and no docks, slipways or repair facilities existed (Mason et al., 1945, 269).

The country's second port, Vlorë, exported 58 000 tonnes and 9000 head of cattle in 1937. A significant proportion of traffic comprised crude oil piped from the inland field near Patos and destined mostly for Italy as aviation fuel, and asphalt bitumen. Vlorë's sheltered bay was capable of accommodating some twenty 10 000-tonne vessels. Both Durrës and Vlorë were relatively well served by inland roads. This was less the case with the smaller southern port of Sarandë (Figures 7.3 and 7.4) whose hinterland extended south and eastwards beyond the country's borders, to Ioannina in Greece and Bitolj in Macedonia as well as to Korçë. Both Sarandë and the northern port of Shëngjin were largely concerned with coastwise traffic; the latter, however, had good access to Lezhë and the Tirana – Shkodër road.

In the post-war period, the maritime role of Durrës has been dominant within the country, as oil exports from Vlorë ceased in order to make petroleum available for the rising domestic market. Since the 1960s, Durrës has consistently handled over 80% of the country's merchant seaborne freight (Table 7.1). Further, since the mid-1980s a number of regular ferry services have provided key links with Adriatic neighbours – Bari (9 hours journey), Ancona (20 hours), Koper (22 hours) and Trieste (23 hours) (White et al.,1995). Vlorë (Bërxholi, 1985), handling less than 10% of the country's maritime trade, has acted as the country's major naval

Figure 7.3.   Coastal freighters, Albania. An Albanian coastal freighter approaches the country's third most important port, Sarandë, which lies opposite northern Corfu.

Figure 7.4.   Sarandë, Albania. Several coastal freighters and patrol vessels in Albania's small, yet third most important port, Sarandë. Capacity restrictions are evident.

Table 7.1. Goods handled by Albanian seaports, 1960–90 ('000 tonnes) with percentages of total in italics

	1960	1970	1980	1985	1989	1990
Durrës	1107	2069	2227	2181	2773	2336
	77.5	*87.7*	*81.2*	*84.2*	*84.4*	*82.9*
Vlorë	167	192	273	232	260	241
	11.7	*8.1*	*10.0*	*9.0*	*7.9*	*8.5*
Sarandë	99	50	149	133	179	150
	6.9	*2.1*	*5.4*	*5.1*	*5.4*	*5.3*
Shëngjin	55	48	93	44	75	92
	3.9	*2.0*	*3.4*	*1.7*	*2.3*	*3.3*
Total	1428	2359	2742	2590	3287	2819
	100.0	*100.0*	*100.0*	*100.0*	*100.0*	*100.0*

Sources: KPS (1989, 108), DeS (1991, 270–1) and author's additional calculations.

military base, including an offshore former Soviet submarine base on Sazan island (Heiman, 1964) and an onshore naval academy. Ferries from Brindisi (8 hours), Otranto (8 hours) and Bari ($9\frac{1}{2}$ hours) now call at the port. In 1994/95 the port of Vlorë became notorious as a 'trampoline' for bouncing illegal immigrants – Kurds, Chinese and Bangladeshis as well as Albanians – through the EU's Italian back door. An average monthly rate of 5000 smuggled migrants are reckoned by Italian police to take this route (Anon., 1995).

With a population of 120 000, Durrës has become the country's second largest city, focal point of Albania's rail system, and centre of commerce (Bërxholi, 1986). For a short time after independence it was the country's capital. While the port has retained a pre-eminence in external maritime trade, in international terms it is very restricted in size, draught and technology, with a port basin of only 67 hectares and quayside cranes with a maximum lifting capacity of 15 tonnes. Critically, Durrës has been unable to handle container ships, and has had the unenviable reputation of being one of the least efficient ports in Europe.

Nearby Thessaloníki (Hoffman, 1968; Papayannopoulos, 1982; Darques, 1992; see also Chapter 9), by contrast, has been ready and able to capture post-communist Albanian trade. Under the Turks, the port was second only to Constantinople as the major commercial outlet for the Balkans. Access through the mountainous interior into Serbia via the Vardar valley, gave it the potential to be the major sea port for the Danube basin. When, in 1912, the city passed from Ottoman to Greek rule, much of its hinterland was annexed by Serbia, immediately increasing the port's potential, since Serbia saw Thessaloníki as its principal access to seaborne commerce. Developments were, however, cut short by the outbreak of war. More recently, both Bulgaria and Serbia have again made use of the port, although United Nations sanctions against Serbia inhibited this role, as did the Greek unilateral embargo imposed on the former Yugoslav

Republic of Macedonia, curtailing an estimated $105 million worth of annual trade between the two countries (Smith, 1994). In recent years, in the face of poor Greek–Albanian relations (Hall, 1996), several Albanian importers have opted to unload seaborne goods at Thessaloníki and bring them into Albania by road. Albania has some 7500 km of road, although only 40% of this total is asphalted.

Plans for change

In 1992 upgrading of Durrës port began with the construction of a new 15 000 square metre ferry quay to permit three passenger vessels to be in port at any one time. Subsequently, a long-term low-interest credit of $8.7 million was secured from Kuwait to assist the port's redevelopment. Beset by port infrastructures requiring comprehensive modernisation and elaboration, Durrës has also been faced with relatively poor landward transport and communications links, despite the cityport being the hub of the country's road and rail systems and situated only 50 km from Albania's international airport.

As the focal point of the 720-km Albanian rail system, Durrës contains the main workshops and marshalling yards as well as a small port railway. The city's terminal station acts as the apex of the passenger network: services operate inland to Tirana, to Shkodër in the north, Pogradec in the south-east (Figure 7.5) and Vlorë in the south-west. Economic plans in the 1980s gave priority to doubling the capacity of the rail network, including completion of the first international link, from Shkodër to the then Titograd (Podgorica) in Montenegro, as well as completing the line to Vlorë to provide a rail link between Albania's two major ports (Kromidha and Konduri, 1984).

Only after much controversy between the two partners was the international rail link finally inaugurated in 1986 (Hall, 1984, 1985, 1987). The line has been disused and the Albanians are seeking compensation for the adverse economic impact of UN sanctions on Serbia and Montenegro. Topographic conditions in the country are far from ideal for railway development, with upland barriers and steep gradients in the centre and east of the country and formerly swampy lowlands in the west. Unfortunately, Albania's hydro-electric power resources have yet to see any electrification of the system, diesel power having monopolised traction for most of the railways' existence. In their waning months the communists signed a protocol with the Italian government which included the provision for investment in electrification. For the communists' aborted 1991–95 ninth five-year plan, an extension of the railway system to the Greek border had also been mooted.

Following a highly disruptive period of social unrest in the early 1990s, 63 passenger coaches were donated by Italy, followed by rails and 50 000 sleepers: passenger services at least were consequently restored during 1992. The Railway Directorate estimated that $200 million was needed to renovate the system, and plans for a 120-km link from Pogradec to Florina in Greece via Korçë, at a cost of $260 million, were again discussed, with the possibility of finance from European

Figure 7.5. Rail transport in Albania. Part of the existing Albanian main-line railway passing through the centre of the country from the Adriatic port of Durrës to the town of Pogradec close to the Macedonian border. Just to the north of Pogradec, by Lake Ohrid, is the village of Lin, a corner of which is the backdrop for this passenger train on its journey to the coast. As part of the axis for a potential international route, the line is currently single-tracked for most of its length.

Community development funds (Sarbutt, 1992). The latter agreed to finance a modernisation of the railway's telecommunications system and to provide funding for other vital supplies, while Italian State Railways would provide retraining programmes for Albanian rail officials.

About two-thirds of both freight and passenger traffic has been carried on the country's road network since the mid-1980s (Table 7.2). Yet apart from the recently completed 32-km Durrës–Tirana highway, all of the country's main roads are badly rutted, virtually unlit and bedevilled by slow-moving erratic vehicles and animal transport. Since private vehicles were legalised in 1991, a rapid increase in mostly poorly maintained, ageing second-hand motor vehicles – 17 000 were imported in 1992 – has added substantial problems to the already inadequate road system, and abandoned wrecks soon became a major driving hazard (McDowall, 1991) (Figure 7.6). However, during 1993 a regional enterprise was being established to maintain the national road system. Funding for this and for new and improved roads within the country was in the form of World Bank credits, $18 million of which was coming from the International Development Association and $8 million from Kuwaiti sources. In 1994, however, Ministry of Construction spending on road development amounted to just $790 000 on major routes and $150 000 on urban road improvement (*Radio Tirana*, 31 December 1994).

Table 7.2.   Albanian freight transport: the modal split, 1950–90

	1950	1960	1970	1980	1985	1989	1990
Road	1499	10278	34269	62021	75796	82815	75744
	57	329	776	1302	1221	1304	1195
	75	81	81	72	65	65	66
Rail	239	875	2324	5806	8114	8048	6646
	9	55	160	477	605	674	584
	11	14	17	26	34	33	32
Coastwise/inland	101	124	97	225	237	411	377
shipping	11	20	17	34	30	36	34
	14	5	2	2	2	2	2
Totals	1839	11277	36690	68052	84147	91274	82767
	76	404	953	1813	1856	2014	1813
	100	100	100	100	100	100	100

Key
1499   in thousand tonnes
  57   in million tonne/kilometres
  75   % of goods transported domestically in tonne/kilometres
Sources: KPS (1989, 105–6), DeS (1991, 264–7) and author's additional calculations.

Figure 7.6.   Road transport in Albania. A series of mountain hairpin bends which are not uncharacteristic of most roads passing through Albania, and indeed through much of the southern Balkans, exemplifying some of the physical constraints on landward distribution from Albanian ports. Here a long-distance bus passes a disabled lorry at 600 metres elevation in southern Albania.

From 1991, international courier companies began to establish themselves within the country, aiding communications links with the outside world particularly for the growing business community. Reflecting the reduction in technological and political impediments to flights into and out of Tirana, the carrier DHL opened an office in Albania in July 1991. An exclusive representative agency agreement with Albtransport provided a staff of four and an office in the centre of Tirana, with DHL supplying equipment, transport and staff training. By the following year, the company was shipping 200–500 packages per month into Albania, and about 80 out of the country (Kobylka, 1992).

Opportunities?

Somewhat belatedly, upgrading of Albania's infrastructure became a major priority of the 1993 economic programme, assisted by an $80-million World Bank facility and additional assistance from the European Community and the EBRD (European Bank for Reconstruction and Development). Two elements which were being looked to for relatively quick results were the redevelopment of Durrës port and Rinas international airport (Keay, 1993). Generally, however, negotiations between the government, international financial institutions and possible private investors moved at a very slow pace: even after IMF support in mid-1992, bureaucratic obstacles and the passing of inconsistent legislation had the effect of dampening overseas enthusiasm for inward investment.

None the less, the role of Durrës is to be enhanced considerably if a 19th-century dream is finally realised: major road and rail links from Bulgaria and Macedonia to the Albanian coast linking the Adriatic with the Black Sea. Political change in Albania coupled with subsequent events in Yugoslavia have substantially encouraged such development. The possibility that Albanian ports, desperately in need of new infrastructure and technology, might be opened up as free trade zone areas at one end of this link, was initially signalled in the summer of 1990 when the private export–import company Makimport of Skopje was offered talks on the use of Vlorë's port facilities. Subsequent Macedonian interest was also shown in establishing Durrës as a free port. In 1991, the Skopje railway enterprise of Macedonia completed plans for the construction of 121 km of track linking the Bulgarian, Macedonian and Albanian rail systems.

By improving communications, such a line would considerably shorten the distance for shipping goods from Italy to Eastern Europe and the Middle East, and of course, would circumvent the problems of Yugoslavia. Durrës, the line's Adriatic terminus, would stand to gain considerable advantage from this. The cost of the rail link has been estimated at $350 million.

For some years there had been plans on the international drawing board to create a cross-border highway network from Albania as part of the Trans-European Motorway Project. The main routes would be west–east, Durrës to Pogradec, and on into Macedonia, Greece and the Black Sea, and north–south, from Shkodër to the Greek border, probably linking Tirana, Berat and

Gjirokastër. A 1000-km highway from Durrës to Istanbul could reduce a 30-hour journey by two-thirds. Albania was looking to international development aid programmes to fund the estimated $500 million cost of its stretch of the road, while Macedonia was hoping to attract private investment for its likely $280 million commitment.

With the Yugoslav situation worsening, in May 1993, Albanian, Bulgarian and FYR Macedonian transport ministries agreed to go ahead with Black Sea–Adriatic high-speed road and rail links, air routes and a digital telecoms link. Italy and Turkey would join the project later. Financing from the EBRD was envisaged. Subsequently, the EU made ECU 100 million in PHARE project aid available to the adjacent countries most affected by UN sanctions against Serbia/Montenegro (Albania, Bulgaria, Hungary, the FYR of Macedonia, and Romania). These funds were earmarked mainly for the construction of alternative roads through the region, not least the Adriatic–Black Sea link. Russia was also supportive of a Durrës–Skopje–Sofia(–Varna/–Istanbul) link since it would give the country access to the Adriatic via Romania and Bulgaria. Further, on the establishment of a partnership between 38 Albanian transport and communications enterprises and the Kuhne Nagel International group, the German partners emphasised that they considered Durrës to be the best port linking (Western) Europe with Ukraine, Romania and Bulgaria, and acknowledged the potential for overland transport requirements from Albania, promoting their new Transalbania joint venture.

As part of a wider package of agreements on cooperation in transport and communications with Croatia, the Albanian Posts and Telecommunications Department also agreed to participate in a German–Croatian–Greek fibre-optic cable project, whereby a submarine cable would be laid from Dubrovnik to Durrës and then on to Corfu (Albanian Telegraph Agency (ATA, Tirana) 16 March 1993; EIU, 1993). These telecoms developments symbolised Albania's new self-perceived role of wishing to act as a neutral east–west, north–south communications and cultural bridge. Agreements with Bulgaria, the FYR of Macedonia, Italy and Turkey to establish the Trans Balkan line telecoms bridge entail the construction and maintenance of a 1000-km long digital system of optical fibre cables and digital radio relay systems to connect the five participating countries to each other and to Western Europe, again avoiding the central Yugoslav vortex.

IMPLICATIONS

'Deflected infrastructure'

With the short- to medium-term requirement of the international community to circumvent Serbia, Albania appears to be on the threshold of experiencing a major transport and communications upgrading programme. However, the question needs to be posed that in supporting an Adriatic–Black Sea land and air bridge via FYR of Macedonia and Bulgaria, to what extent will Albania's transport infrastructural upgrading – port redevelopment, international road, rail and telecoms links, international airport upgrading and overflying rights – be relevant and appropriate to the immediate needs of Albania itself?

Second, what is likely to happen to those developments when Serbia is readmitted to the world community and international links via the Sava–Morava axis are restored and improved? While funding agencies are apparently willing to support what may well be a short-term palliative for unblocking an international bottleneck, on previous performance there is little evidence to suggest that significant inward investment will be attracted to generate permanent employment opportunities for Albanians. Cumulative FDI for 1990–93 amounted to just $39 million, representing a mere $18 per capita (Keay, 1995). Further, economic leakages from major infrastructural development are likely to be substantial: for example, the Albanian construction industry could not even be entrusted by Austrian developers to build a new Tirana hotel – Slovenian sub-contractors were brought in.

Unrealistic perceptions?

Albania will seize with open arms any offers of infrastructural upgrading, given the poor state of the existing fabric. Such upgrading will inevitably be perceived by Albanians as an opportunity to better attract inward investment and thereby secure longer-term economic development. But if the 'deflected infrastructure' is being established as an expedient transit route, as an international palliative which could very easily be dispensed with, the result for Albania could be (a) substantial environmental damage, (b) inappropriate infrastructure (and developments which focus on Durrës rather than on the capital Tirana), and (c) falsely raised hopes of economic opportunities.

Competing influences

A major debate within Albania continues to surround the direction which the country's development path should take. This has both economic and cultural as well as political dimensions. Economic assistance is being offered to the country from diverse sources. It would seem that many Albanians, while recognising their predominantly Islamic background, wish to be part of 'Europe', and as such look westwards for their economic and political role models (*Flash – Albania* (Tirana), March 1993; Hall, 1994, 268–9). But with the extreme sensitivity now attached to cultural and particularly religious attributes in the region – not least in Bosnia and Kosovo – the role of Albania's 'eastern' historic evolution, however pragmatic the adoption of Islam may have been under the Ottomans, is today not insignificant. Aid and assistance from the Islamic world, notably from Turkey and the Gulf states, entrepreneurial forays by Greeks, substantial assistance from Italy, institutional funding from the major Western banks, as well as hectic proselytising activity within Albania from well and lesser-known competing religious faiths, render Albania again a stage for competing economic and cultural influences to extend and counteract their spheres of influence. The nature of support for the country's transport infrastructure development and bridgehead role may be seen as one element of that competition.

CONCLUSIONS

This chapter has looked briefly at some of the implications of regional restructuring in the Balkans for transport development there. With rapid change and upheaval being experienced, uncertainty will continue to play a dominant role in any analysis of economic (and morphological) restructuring processes (Hall, 1993a; Hall and Danta, 1996).

That Albania was hermetically sealed from the rest of the Balkans for several post-war decades presented a thoroughly alien concept to the region. Just before the Second World War, Albania's geopolitical role over the previous 2000 years had been interpreted as that of a 'gateway' (Stadtmüller, 1937–38), a notion revived in political geography as the 'new world disorder' was making itself felt in the region (Cohen, 1990). Indeed, a potential west–east link role for Albania was evident even before the Yugoslav wars of succession brought conflict to the region. The additional need for at least short- to medium-term avoiding strategies for transport development focuses the spotlight even more intensely on Albania's key role. The country itself is hampered by comprehensively poor infrastructures, continuing political uncertainty and understandable hesitance on the part of potential inward investors. The likelihood of an international land and air bridge requiring infrastructural development which may have only a short-term international relevance and which may not necessarily be the most appropriate for Albanian needs, raises important questions for national and regional development strategies, the role of aid and investment, and the nature of competing spheres of influence in the Balkans.

REFERENCES

Anon. (1995), '. . . as much as Albanian smugglers', *The Economist*, 1 April.
Bërxholi, A. (1985), *Vlora and its environs* (Tirana: 8 Nëntori).
Bërxholi, A. (1986), *Durrës and its environs* (Tirana: 8 Nëntori).
Cohen, S.B. (1990), 'The world geopolitical system in retrospect and prospect', *Journal of Geography* 89, 2–12.
Cvijić, J. (1918) 'The zones of civilization of the Balkan peninsula', *Geographical Review* 5, 470–82.
Darques, R. (1992), 'L'espace urbain de Salonique: de la ville Ottomane à la metropole Grecque moderne', *Bulletin de la Société de Geographie de Marseille* 91(20), 42–54.
DeS (Drejtoria e Statistikës) (1991), *Vjetari statistikor i Shqipërise* (Tirana: DeS).
EIU (Economist Intelligence Unit) (1993), *Romania, Bulgaria, Albania: country report No. 2 1993* (London: EIU).
Hall, D.R. (1984), 'Albania's growing railway network', *Geography* 69(4), 263–5.

Hall, D.R. (1985), 'Problems and possibilities of an Albanian–Yugoslav rail link', in Ambler, J. et al. (eds), Soviet and East European transport problems (London: Croom Helm), 206–20.

Hall, D.R. (1987), 'Albania's transport cooperation with her neighbours', in Tismer, J.F. et al. (eds), Transport and economic development – Soviet Union and Eastern Europe (Berlin: Duncker and Humblot), 379–99.

Hall, D.R. (1992), 'East European seaports in a restructured Europe', in Hoyle, B.S. and Pinder, D.A. (eds), European port cities in transition (London: Belhaven), 98–115.

Hall, D.R. (1993a), 'Impacts of economic and political transition on the transport geography of Central and Eastern Europe', Journal of Transport Geography 1(1), 20–35.

Hall, D.R. (ed.) (1993b), Transport and economic development in the new Central and Eastern Europe (London: Belhaven).

Hall, D.R. (1994), Albania and the Albanians (London: Pinter).

Hall, D.R. (1996), 'Recent developments in Greek–Albanian relations', Mediterranean Politics 2.

Hall, D.R. and Danta, D.A. (eds) (1996), Reconstructing the Balkans (London: Wiley).

Heiman, L. (1964), 'Peking's Adriatic stronghold', East Europe 12, 15–16.

Hoffman, G.W. (1968), 'Thessaloniki: the impact of a changing hinterland', East European Quarterly 2(1), 1–27.

Hösch, E. (1972), The Balkans (London: Faber).

Keay, J. (1993), 'Life after Hoxha', EuroBusiness 1(3), 42–6.

Keay, J. (1995), 'Unrewarded eastern promise', EuroBusiness 2(10), 30–3.

Kobylka, J. (1992), 'DHL brings some help to Albanian mail', Business Eastern Europe 21(28), 459–69.

KPS (Komisioni i Planit të Shtetit, Drejtoria e Statistikës) (1989), Vjetari statistikor i R.P.S. të Shqipërisë (Tirana: KPS).

Kromidha, T. and Konduri, P. (1984), 'Achievements and development of transport', Albania Today 78, 33–5.

Mason, K. et al. (1945), Albania (London: Naval Intelligence Division).

McDowall, L. (1991), 'Albania learns the art of wrecking', New Statesman and Society, 13 December.

O'Sullivan, J. (1972), The Egnatian Way (Newton Abbot: David & Charles).

Papayannopoulos, A. (1982), History of Thessaloniki (Athens: Rekos).

Raven, S. (1993), 'The road to empire', Geographical Magazine 65(6), 21–4.

Sarbutt, G. (1992), 'Rail crisis in Albania', Modern Railways 48(11), 622–3.

Singleton, F.B. and Wilson, J. (1977) 'The Belgrade–Bar railway', Geography 62(2), 121–5.

Smith, H. (1994), 'Athens shuts Balkan gateway to Macedonians', The Guardian, 3 March.

Stadtmüller, G. (1937–38), 'Landschaft und geschichte in Albanisch-epirotischen raum', Revue Internationale des Etudes Balkaniques 3, 345–70.

White, L.L., Dawson, P. and Dawson, A. (1995), Albania: a guide and illustrated journal (Chalfont St Peter: Bradt).

Wilson, O. (1971), 'The Belgrade–Bar railroad: an essay in economic and political geography', in Hoffman, G.W. (ed.), Eastern Europe: essays in geographical problems (London: Methuen), 305–93.

8 Patras and its Hinterland: Cityport Development and Regional Change in 19th-Century Greece

ELENA FRANGAKIS-SYRETT
Queens College and the Graduate Center, City University of New York, USA
MALCOLM WAGSTAFF
Department of Geography, University of Southampton, UK

The close relations which exist between the cityport and its hinterland are symbiotic in character. The city flourishes on the activity in its port and that, in turn, serves the requirements of the hinterland. Economic growth in the hinterland is likely to generate not only increased activity in the port but also population growth in the city. Further developments in the cargo-handling facilities of the port can be postulated, while an expansion of both the cityport's infrastructure and also of its built-up area are probable. Close chronological phasing is likely in the physical development of the cityport, on the one hand, and economic growth in the hinterland, on the other. The character of port activity in terms, for example, of the types of cargo handled and the seasonality of shipping movements will reflect the nature of the economy in the hinterland. Changes in the regional economy imply changes in port activity, as well as adjustments in both the built form and the socio-economic structure of the city.

These relationships are explored in this chapter through an examination of Patras and its hinterland during the period *c.* 1830 to 1907. The link between the two is the trade in currants. Currant production expanded in the hinterland throughout much of the 19th century. The labour demands thus generated, together with the ready availability of land, attracted population to the region and the city. The export of currants dominated not only the activity of the port but also the economic and social structure of the city. When the demand for currants collapsed towards the end of the century, Patras took on a new role.

THE CITYPORT OF PATRAS

Patras lies on the north-west coast of the Peloponnese, just over 8 km west of the narrow entrance to the Gulf of Corinth or Lepanto where a curve in the coastline

Cityports, Coastal Zones and Regional Change. Edited by Brian Hoyle.
© 1996 John Wiley & Sons Ltd.

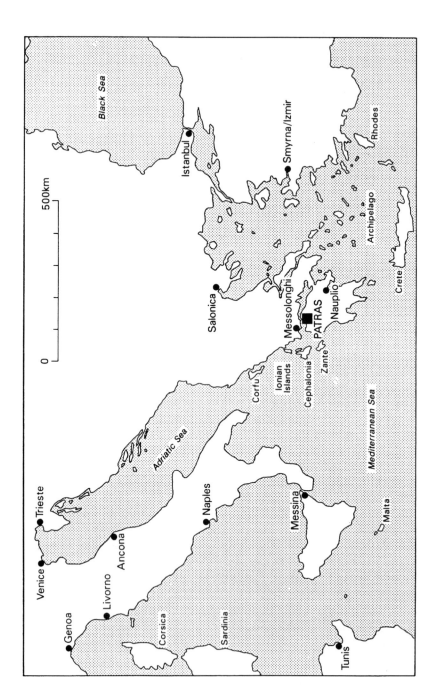

Figure 8.1. Location of Patras, Greece

provided a little shelter against the wind (Figure 8.1). As the English traveller, Edward Dodwell, observed at the beginning of the 19th century, it was well placed to control the trade of the Gulf (Dodwell, 1819, 119). Location also favoured the development of Patras as a focus for trade with the nearby Ionian Islands and the neighbouring coastal areas of Greece and Albania. To the north it enjoyed access to the ports of the Adriatic Sea, while to the west it was open to Sicily, western Italy, southern France, eastern Spain and the Atlantic world beyond the Straits of Gibraltar (Leake, 1830, II; Philippson, 1892).

Once the Greek War of Independence was over (1830), Patras resumed its position as a leading Greek port, particularly in the export trade. Within a decade Austrian and British steam-shipping lines used it and by the second half of the 19th century steamers called there at least fortnightly, if not weekly, from ports in the western and eastern Mediterranean, carrying passengers as well as goods. Direct sailings to the USA were established by the end of the century. Throughout the 19th century the port of Patras enjoyed a strong commercial link with Britain, based very largely on the export of currants from its extensive hinterland which covered the western Peloponnese, the mainland of central Greece and the Ionian Islands (FO 32/81, Crowe, Patras, 2 February 1838; FO 286/280, Annual Trade Report for 1871). In addition, Patras was a major importer of British textiles for distribution throughout the same territories (FO 32/73, Crowe, Patras, 18 January 1837; FO 286/384, Wood, Patras, 17 January 1887).

Figure 8.2. Patras, Greece, in 1862. (Source: *Illustrated London News*, Vol. XVI, no. 1174, 15 November 1862, p. 509)

The core of the early 19th-century town lay below the citadel hill, across the slopes to the south-west, and a short distance from the sea (Pouqueville, 1813; Leblanc, 1829). The hills merge into a narrow coastal plain where a significant extension of the town had developed behind the shoreline before 1821 (Leblanc, 1829) and where much of the urban growth took place later in the 19th century. The coastal plain extends southward and westward into a tract of dunes, lagoons and swamps in the neighbourhood of the hill of Mavrovouni and Cape Araxos, some 14–15 km from the town. The area is dominated by the detached ridge of Mt Skollis (965 m). To the north of Patras an area of marsh remained largely unreclaimed throughout the 19th century (Dodwell, 1819; Admiralty Chart, 1864, 1895), though much of the narrow plain in this direction consists of alluvial fans spreading out from rugged hills to the east. Mt Panakhaikon (1926 m) overlooks the area. The hills of folded shales, sandstones and conglomerates run south and west, their contact with the coastal plain marked by a distinct break of slope at about 400 m. They rise inland to a high escarpment which marks the effective boundary of the Patras administrative district (*Eparkhia*) (Figure 8.2). Even in the late 19th century, these hills were covered with mixed woodland (oak, fir, Aleppo pine), though there were extensive areas of cultivation as well (Philippson, 1895).

POPULATION GROWTH: THE TOWN

The commercial and demographic growth of Patras started in the late 18th century (Wagstaff and Frangakis-Syrett, 1992), but it was savagely cut back by the Greek War of Independence (1821–30) (SP 105/139, Green, Patras, 7 April 1821; SP 105/140, Green, Patras, 28 October 1822). The town was fought across and, consequently, emerged from the war as a wreck. Its population was estimated at 437 families or perhaps 2076 individuals (Commission Scientifique, 1835), considerably lower than pre-war estimates which had been as high as 10 000 (e.g. Leake, 1830, II, 140). Demographic recovery, however, appears to have been relatively rapid.

By the late 1830s the population was estimated at 6–8000 (*Μπακουινακης*, 1988, 50, table 2). In 1848 the *demos* (town) of Patras reported a population of 15 400 (*Μπακουινακης*, 1988, 50, table 2), most of whom must have lived in the town itself. Patras was the third most populous *demos* in Greece, as then constituted. Athens, the capital, contained 26 256 people and Ermoupolis (Syra), the leading port, 19 410 (*Χουλιαρακη*, 1974). In 1856 Patras town contained 15 131 people (Table 8.1), a figure equivalent to 79% of the population of the *demos* and 41.2% of the *eparkhia* (Table 8.2). The *demos* of Patras, with a population of 19 138, now ranked second to Athens (33 136) and was ahead of Ermoupolis (18 830). It retained this position until the end of the century. At the 1907 census, the *demos* of Patras (48 938) again ranked third, behind Piraeus (71 505) and Athens (167 479). Throughout the century, though, its share of the national population remained roughly constant at around 2.0%. This is remarkable given the substantial boosts to the national population which took

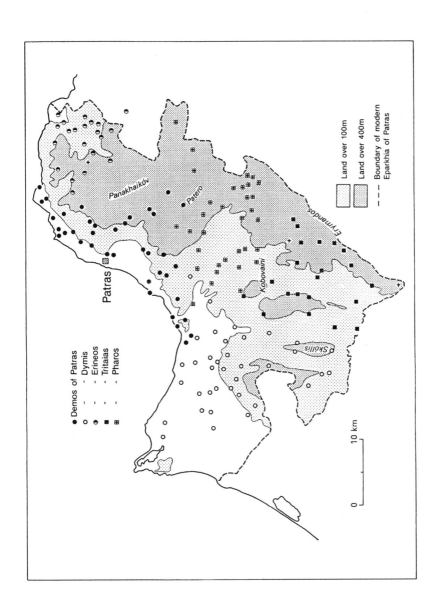

Figure 8.3. The *Eparkhia* of Patras showing relief and the pattern of *oikismi* in 1889. (*Χουλιαράκη*, 1974)

Table 8.1.   Population change in the Patras district, 1856–1907

Year	Town		Eparkhia (excluding Patras)	
	Population	Change (%/annum)	Population	Change (%/annum)
1856	15 131		21 572	
		4.1		1.5
1861	18 342		23 147	
		−1.0		3.2
1870	16 641		29 886	
		3.4		0.7
1879	25 494		31 679	
		3.1		2.0
1889	33 529		38 004	
		1.6		0.9
1896	37 985		40 458	
		−0.1		1.2
1907	37 724		45 754	

Source: Χουλιαρακη (1974).

Table 8.2.   Proportion of the local population resident in the cityport of Patras, 1856–1907

Year	Percentages		
	of *demos* population	of *eparkhia* population	of *nomos* population
1856	79.0	41.2	16.8
1861	79.7	44.2	19.9
1870	63.5	35.8	16.9
1879	74.5	44.6	22.0
1889	74.5	46.9	25.6
1896	75.7	54.2	26.2
1907	72.6	55.2	25.0

Source: Χουλιαρακη (1974).

place with the acquisition of the Ionian Islands (1864) and Thessaly (1881) (Χουλιαρακη, 1974).

   Annualised rates of growth for the town in the period 1856 to 1907 are given in Table 8.1. They were particularly high in the 1850s, the 1870s and 1880s. In these years the population of the town grew at a slightly greater rate (an average of 2.9% per annum) than that of the *eparkhia* (an average of 2.5% per annum). The difference is explained by immigration to the developing port city. The years 1878 to 1893 particularly were the boom period for the export of currants and demand for labour was high. With the exception of 1870, there was a rising trend in the proportion of the local population reportedly resident in Patras itself (Table 8.2). By 1907, 55.2% of the population of the *eparkhia* lived in the city and 25.0% of the population of the *nomos* (province) of Akhaia, a proportion actually a little lower than in 1896. Rapid population growth associated with population concentration, as well as phases of population decline and stagnation, must be related to local economic conditions.

As population grew, so the town itself expanded. The irregular layout of the pre-war settlement was replaced by two rectangular grids, characterised by wide streets and large squares (Figure 8.4). One of them lay in the area south-west of the citadel (which became an army barracks and prison). The other was spread across the coastal plain between the citadel and the port. The 'upper' and 'lower' towns of the early 19th century had merged well before the beginning of the 20th century and surviving open spaces in the built-up area were gradually filled with buildings. The 'lower' town opened on to an extensive waterfront. Finger quays, a mole and breakwater improved the ship-handling facilities here during the 1880s (Figure 8.4), though progress was slow (Τζετζος, 1885; Κωστοπουλος, 1902).

The railway from Piraeus and Athens arrived at the northern end of the waterfront in 1887 and was provided with its own quay, while the line from Pyrgos and the western Peloponnese reached the southern end of the waterfront in 1890 and also had its quay (Struck, 1902). The two stations became the crystallisation points for further development of the built-up area in the 1890s. They were subsequently linked across the waterfront and a completely new station was built. Hotels appeared nearby and shortly after the beginning of the 20th century the city's amenities included a theatre, as well as post office, law courts, and the seats of the provincial governor and the archbishop (Baedeker, 1909). The National Bank of Greece, established in 1841, had a branch there from 1846 and branches of the British-dominated Commercial and Ionian Banks were also established in the city from the 1840s.

POPULATION GROWTH: THE COUNTRYSIDE

Population in the *eparkhia* of Patras (excluding the city) rose from an estimated 13 572 *c.* 1830 (Commission Scientifique, 1835) to 45 754 in 1907, an overall increase of 237.1% (Table 8.1). The annualised rates of increase were low, except between 1861 and 1870, and again between 1879 and 1889. Population growth was probably accompanied by the development of some new settlements or the reoccupation of sites which had lain deserted for a long time. Unfortunately, the available source material does not allow precise numbers to be given.

The major increase in rural settlements probably took place in the early part of the century. This is suggested by the increase from 127 settlements named *c.* 1830 (Commission Scientifique, 1835) to 152 *oikismi* (registration units) used

128

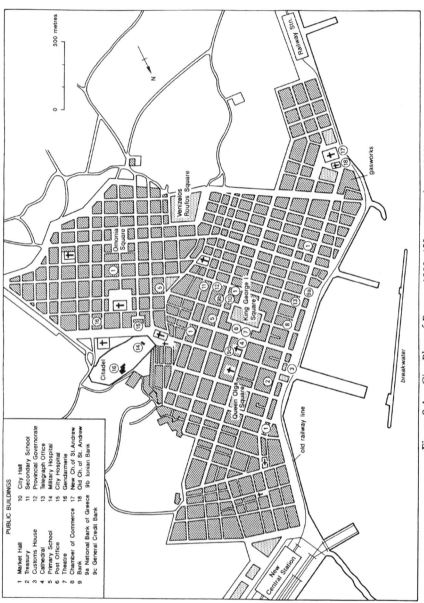

PUBLIC BUILDINGS

1	Market Hall	10	City Hall
2	Treasury	11	Secondary School
3	Customs House	12	Provincial Governorate
4	Cathedral	13	Telegraph Office
5	Primary School	14	Military Hospital
6	Post Office	15	City Hospital
7	Theatre	16	Gendarmerie
8	Chamber of Commerce	17	New Ch. of St. Andrew
9	Bank	18	Old Ch. of St. Andrew
9a National Bank of Greece	9b Ionian Bank
9c General Credit Bank

Figure 8.4. City Plan of Patras, 1902. (*Κωστοπουλος*, 1902)

by the 1879 census (Χουλιαρακη, 1974). A later expansion in settlement numbers is also possible. Between 1889 and 1896, that is, towards the end of the boom years for currant production, the number of *oikismi* rose from 155 to 169 (Χουλιαρακη, 1974), recognition perhaps of sufficient population growth to require separate designation of new *oikismi* (but see note 1). Certainly, the mean population of *oikismi* grew between 1879 and 1907, from 236.7 to 285.1 individuals (Χουλιαρακη, 1974).

How are these changes to be explained? Labour shortage was acute in the Patras region throughout the century and resulted in the migration of people from the surrounding districts, as well as the extensive use of seasonal labour, including people from the Ionian Islands and the mountainous interior of the Peloponnese (FO 286/262, Ongley, Patras, 18 October 1869). In 1876, we are told by Consul Wood, the Greek Government created a totally new settlement halfway between Patras and Pyrgos and settled Albanian-speaking Greek Orthodox farmers from southern Albania (FO 286/308, Wood, Patras, 18 January 1877). Farmers were encouraged to come to the area with free grants of 10 acres (4 ha) of land to each family, a considerable endowment when currant-producing plots varied between 2.5 and 15 acres in size.

Another part of the process of settlement genesis was observed by the young German geographer, Alfred Philippson (1892). This was the growing permanency of low-lying settlements (*kalyvia*: 'huts') used on a seasonal basis by people from settlements at higher elevations, chiefly for the wintering of their sheep and goats. The main incentives for people to abandon the upper settlements were less the rising value of land below about 335 m, the upper limit for the successful cultivation of currant vines (Naval Intelligence, 1944, II), than the opportunity to deploy family labour productively and the ready availability of capital, both local and foreign (Franghiadis, 1990). The stimulus was the expanding trade in currants (Figure 8.5). This peaked in 1891. Although the increase in currant exports ended then, the benefits which flowed from the expansion of currant cultivation and the growth of exports may well have carried population growth in the region, as at Patras itself, beyond the onset of economic crisis. The crisis took time to register locally, despite the bankruptcy of the Greek state in 1893, itself largely a consequence of the currant crisis.

THE CURRANT TRADE

Philippson told his readers in 1892 that the importance of the cityport of Patras lay exclusively in the expansion of currant production in its vicinity. The growth of currant production is shown in Figure 8.6. In 1893 the port of Patras handled 39.2% of all Greek exports by value and 70.5% of the goods exported through Patras were, by value, currants (Υπουργειου Οικονομικων, Γραφειο Στατιστικης, 1894). Even the industrial activities of the town were affected by the dominance of the currant trade. In 1876, for example, two steam mills were grinding sulphur to spray onto the currant vines, while a third was sawing the

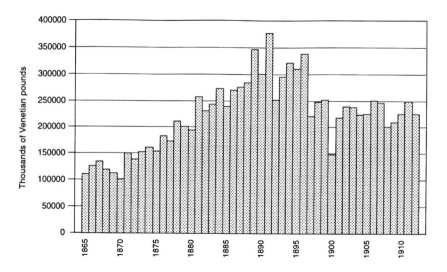

Figure 8.5. Total currant exports from Greece, 1865–1912. (Πιζανιας, 1988, Tables I and II, pp. 128–33)

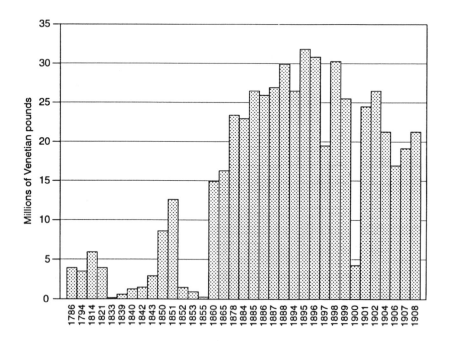

Figure 8.6. Production of currants in the Patras District, 1786–1908. (Franghiadis, 1990, Appendix IV)

planks necessary for making boxes to pack the currants (FO 286/308, Report of 1877 for 1876). The improvements to port installations in the 1880s, already mentioned, were mainly intended to handle the currant trade.

Down to 1877 the major driving force in the expansion of currant production was demand from northern Europe and especially Britain (Coumoundouros, 1859, iv; FO 286/280, Annual Trade Report for 1871), where currants were used in cakes, puddings and pies. Immediately following the War of Independence, it was largely a question of resuscitating the plantations created in the late 18th – early 19th century (Wagstaff and Frangakis-Syrett, 1992) which had been destroyed or neglected because of the fighting (Strong, 1842).

The first currants must have reached the international market in the late 1830s, as vines planted again after the end of the war would have required at least seven years to come to fruition (Strong, 1842). Production, at that time mostly concentrated in the north-western part of the Peloponnese near Patras, did not reach pre-war levels until the 1840s (Figure 8.6), though the area under currants was reported to be 12 556 stremmata (in the early 1840s) and new plantations were allegedly being created at the rate of about 1000 stremmata per annum (Strong, 1842, 176). The average production of currants per stremma (approximately one-quarter of an acre) was as much as 500 Venetian lbs (see note 2) and in some areas much more (FO 286/262, Ongley, Patras, 23 July 1869). Production fell in the early 1850s (Figure 8.6) as a result of very cold weather and white rot on the vines (Coumoundouros, 1859, v; Kipps, 1930–33). In 1853, 4800 tons were unfit for consumption and remained unsold (FO 32/210, Wood, Patras, 24 January 1853). Moreover, this was the second year running that a bad harvest had occurred (FO 32/210, Wood, Patras, 19 August 1853). A combination of good weather and spraying with copper sulphate, which dealt with the problem of disease in the 1850s, led to an increase in currant production. Although vineyards could still be attacked by disease, as the case of *Peronosporos* (a species of white blight) in the 1890s shows, it was not allowed to cause the devastation experienced in the 1850s.

In 1858 the Greek Government took various steps to promote currant cultivation (FO 286/262, Ongley, Patras, 23 July 1869) and, by 1860, 79 201 888 Venetian lbs (37 625 tons) of currants were being exported from Greece, and the total area of the country under currant vines was estimated at 153 058 stremmata (PPAP, 1865, 234). At the time, currant vineyards were valued at 200 drachmas per stremma, about twice that of ordinary vineyards and 20 times that of the mulberry plantations necessary for silk production (PPAP, 1865). The ready market for currants and the lucrative nature of their production induced further expansion, so that by the late 1860s there were 200 000 stremmata under currant cultivation in the Peloponnese (FO 286/262, Ongley, Patras, 23 July 1869), many of them in the coastal plains of Elis, and total exports from Greece exceeded 126 000 000 Venetian lbs (Figure 8.5).

The expansion of currant cultivation was not just a consequence of market forces. It was, in part, also the result of Government measures in the 1860s which were aimed at increasing the area under cultivation by selling land to small-scale cultivators who paid for it in yearly instalments amounting to 12–15% of their annual production (FO 286/470, Wood, Patras, 14 March 1902). In 1871, the Greek Government passed a further land law granting favourable conditions for the purchase of state land, both arable and wasteland, with the proviso that vines, wheat or other crops would be grown on it. The down-payment of a certain amount per stremma was made according to the quality of the land and subsequent payments were to be spread over a period of 26 years. The measure was successful in as much as 100 000 acres (40 000 ha) were estimated to have been placed under vineyards as a result, 'so that we may look for a great increase in their production when the vines came into full bearing', as Consul Wood observed (FO 281/308, Patras, 18 January 1877; see also Freris, 1986).

Currant production was particularly attractive on small, family-run holdings. Often these were not large enough to support an entire household by subsistence farming, while the occupier frequently had abundant family labour to deploy. A lucrative cash crop was vital to the survival of many small farms. Moreover, the operations involved in currant cultivation are delicate, lack standardisation and require a degree of skill and local knowledge. The peasant family was ideally placed to meet these requirements in the 19th century (Franghiadis, 1990). On the other hand, development of currant plantations required capital. Borrowing was, therefore, essential. The capital, however, had to be invested over a relatively long period (7–10 years) before any return could be expected. Share-cropping was a particularly attractive way to provide it and suited creditors, large landowners and incoming peasant families (Franghiadis, 1990).

From 1878 to 1893 a further massive expansion of currant plantations took place, though mainly away from Patras in the southern Peloponnese (Franghiadis, 1990; Frangakis-Syrett, 1993). The driving force this time was demand from France, though the international market remained generally strong. The reason is that in 1877 phylloxera began to cause serious damage to French vineyards. As local production fell, French wine merchants began to import currants from Greece to make into wine for blending purposes. French imports of currants rose from 7000 tons in 1877 to 70 401 tons in 1889, the peak year, when they exceeded imports to Britain by some 17 000 tons and accounted for 48.9% of the total production of currants (Frangakis-Syrett, 1993). Moreover, as France was buying lower quality currants at a good price, the added demand resulted in an overall increase in the prices for all qualities, including the best quality imported by Britain, largely from the hinterland of Patras, in the northern Peloponnese. Production nearly doubled.

By 1891, however, phylloxera in France had been overcome through grafting American vines on to French stocks. French wine producers demanded protection from foreign imports, including currants. Import duties on currants were raised from 6 francs to 15 francs per 100 kilos immediately, and then to 25 francs in 1894. Greek currants became uncompetitive (FO 286/478, Wood, Patras, 4 March 1903). Prices fell from £15 to £5 per ton, 'a price scarcely covering the cost of cultivation' (Andréadès, 1906; *Encyclopaedia Britannica*, 1926, 435). The Greek Government again took steps to deal with the situation, but it was unable to relieve the sudden poverty of the many small-holders attracted into currant production in the boom years (PPAP, Vol. XC, #3821, Morea, 1906). Social unrest developed as the farmers attacked the wholesalers (and also their creditors) whom they blamed for their situation (Freris, 1986, 24). Seasonal labour migrants from the mountainous interior of the Peloponnese were hit particularly harshly when wages collapsed. Massive emigration, particularly to the United States, was the result (Fairchild, 1911; PPAP, Vol. CXXV, #3556, Annual Series Report, 1906). Patras took on a new role. It became the principal port of emigration for the United States.

CONCLUSION

The population of Patras continued to grow throughout the 20th century and at the last census (1991) the city boasted 172 763 inhabitants (National Statistical Service, 1993), nearly five times the number in 1907. The city remains the third largest in Greece, after Athens and Thessaloniki. Its immediate region, the *nomos* of Akhaia and the *eparkhia* of Patras, did not experience the absolute decline in population which affected most of the Peloponnese, with the exception of the Argolis, in the years 1940–71 (Baxevanis, 1972), though there was significant emigration abroad in the 1950s and 1960s. The region continued to receive immigrants, the cityport in particular. As in the 19th century, they came not only from the immediate hinterland but also from the whole of western Greece, including Epirus and the Ionian Islands. The draw has been more the opportunity of urban work than the availability of cultivable land. Patras now enjoys a much more secure economic base than it appears to have possessed for much of the 19th century. The port is a significant employer, directly or indirectly, for it remains one of the leading ports in the country, with important ferry connections to the neighbouring Italian ports of Brindisi, Ancona and Venice. These connections have been strengthened as a result of Greek membership of the European Union (1980). Much of the city's service industry is connected with the port, in one way or another. Cargoes must still be routed, while business people and tourists need hotels and other services. But Patras now supports a diversity of manufacturing industry, including clothing and footwear, as well as food processing and furniture making.

    Currants are still exported through Patras. However, land use in the immediate hinterland is now much more diverse than it was a hundred years ago. Wheat and olives remain common, but potatoes, spring and summer vegetables, melons, water melons and citrus are very conspicuous in the lower-lying parts. Although the proportion of the cropped area is now 44% of what it was in 1911 (141 579 stremmata in 1982 compared with 324 390 stremmata in 1911) (Ministère de l'Economie Nationale, 1914; National Statistical Service, 1985), currant plantations are still very conspicuous in the landscape, particularly in the higher parts of the hinterland. This is a legacy of the expansion of currant cultivation in the 19th century. Similar testimony to the former importance of currant production and export is found in the port area of the city. Here, empty and abandoned warehouses still carry the names of 19th-century currant merchants.

The influence of Patras on the land use and settlement patterns of its immediate hinterland appears to be less in the 1990s than in the 1890s. The city is provisioned in part from the neighbouring countryside and it is a major bulking centre for agricultural products, including currants, from its immediate hinterland; the processing of agricultural produce is important. But much more than in the last century, Patras is a major regional centre, dominating western and south-central Greece. It now contains over 50% of the population of the *nomos* of Akhaia and nearly 70% of that of its own *eparkhia*. Always of regional rather than just local significance, the port clearly has national significance today. But as the size of its catchment has increased, Patras has become less dependent on its immediate, agricultural hinterland. The symbiotic relationship demonstrated for the 19th century now seems less clear, less direct and more subtle.

ACKNOWLEDGEMENTS

The authors wish to thank the Cartographic Unit of the University of Southampton for the preparation of the maps and graphs, as well as Stella Stravridis for locating the thesis by Franghiadis and for considerable assistance in obtaining Greek statistics on the export of currants.

NOTES

1. The administrative sub-units used to present the census information (*demi, kinotites*) were not subdivided until 1879 and, as a result, individual settlements were not recorded. Accordingly, there is a gap in the information about settlements (classified as *bourgs, villages, hameaux*) between c. 1830, when the French Commission Scientifique named individual places in their enumeration of the population, and 1879 when *oikismi* ('localities') were first used. The names of *oikismi* do not always equate to individual inhabited places (settlements) and it is clear from comparing the lists for the censal years from 1879 to 1907 that there was no consistency about the inclusion of *oikismi* names in the censuses. The appearance of apparently new names in a census (e.g. 1889 compared with 1879) does not necessarily mean that new settlements were established between censuses, any more than the decline in the number of *oikismi* means the actual desertion of settlements (e.g. 1896 compared with 1907).
2. 400 Venetian lbs (*lire*) = 420 English lbs (avoirdupois). Thus, 1.05 English lbs = 1 Venetian pound; 2240 English lbs = 1 ton; c. 2105 Venetian pounds = 1 ton).

REFERENCES

Andréadès, A. (1906), 'The currant crisis in Greece', *Economic Journal* 16, 41–51.
Baedeker, K. (1909), *Greece: handbook for travellers*, 4th edn (Leipzig: Baedeker).
Baxevanis, J. J. (1972), *Economy and population movements in the Peloponnisos of Greece* (Athens: National Centre of Social Research).

Commission Scientifique de Morée (1835), *Relations du voyage de la Commission Scientifique de Morée, Atlas* (Paris and Strasbourg).

Coumoundouros, M[onsieur] (1859), *Tableau général du commerce de la Grèce pendant l'année 1858, dressé et publié par ordre de M. Coumoundouros, Ministre des Finances* (Athens).

Dodwell, E. (1819), *A classical and topographical tour through Greece, during the years 1801, 1805 and 1806* (London: Rodwell and Martin).

Encyclopaedia Britannica (1926), 13th edn (London and New York).

Fairchild, H. P. (1911), *Greek immigration to the United States* (New Haven, Connecticut: Yale University Press).

FO = Foreign Office, General Correspondence, Public Record Office, London.

Frangakis-Syrett, E. (1993), 'The port-city of Patras in the nineteenth century', *Review* 16, 411–33.

Franghiadis, A. (1990), 'Peasant agriculture and export trade: currant viticulture in southern Greece, 1830–1893'. PhD thesis, European University Institute, Florence.

Freris, A. F. (1986), *The Greek economy in the twentieth century* (London and Sydney: Croom Helm).

Kipps, J. (1930–33), 'The Greek currant trade: a study in valorisation under public auspices', *Economic History* 2, 137–52.

Κωστοπουλος, I. xαι, A. (1902), Σχεδιυ Πατρων 1:3000 (Πατρας).

Leake, W. M. (1830), *Travels in the Morea* (London: John Murray).

Leblanc, G. (1829), *Plan de Patras*, Archives historiques, Ministère de la Guerre, Paris. Dépôt de la Guerre, Archives des Cartes, Grèce. R. 18. C. 3. 4. 10. C. 77. Reproduced in Τσαχοπουλος, Π. (1994), 'Οι μαρτυριες περιηγη`των xαι απεσταλμειων, για τις πολεις της Οθωμαυικης Πελοποννησου xαι οι τοπογραφικες αποτιτωσεις της περιοδου, 1828–1836', in Kalligas, H. A. (ed.), *Travellers and officials in the Peloponnese* (Monemvasia: Monemvasiotikos Omilos), 187–222 (216).

Ministère de l'Economie Nationale. Direction de la Statistique (1914), *Recensement Agricole 1911. V. Peloponnese* (Athènes).

Μπακουυακης, N. (1988), *Μια Ελληυικη Πρωτευουσα στου 19° Αιωυα, 1826–1860* (Αθηυα: Εxδοσεις Καστανυωτη).

National Statistical Service (1985), *Agricultural statistics of Greece, Year 1982* (Athens).

National Statistical Service (1993), *Greece in figures, 1992/93*, special issue (Athens).

Naval Intelligence Division (1944), *Geographical handbook series: Greece* (London).

Philippson, A. (1892), *Der Peloponnes* (Berlin: R. Friedlander und sohn).

PPAP = *Parliamentary Papers, Accounts and Papers* (London: HMSO).

(1865) *Consular Reports, Greece*.

(1906) Vol. XC, #3821, *Morea*.

(1906) Vol. CXXV, #3556, *Annual Series, Report*.

Πιζανιας, Π. (1988), *Οικουομηκη Ιστοια της Ελληυικης Σταφιδας* (Αθηυα).

Pouqueville, F.C. (1813), *Travels in the Morea, Albania and other parts of the Ottoman Empire*. Translated from the French by Anne Plumptre (London: H. Colburn).

SP = State Papers

Strong, F. (1842), *Greece as a kingdom, or a statistical description of that country from the arrival of King Otho, in 1833, down to the present time* (London: Longman, Brown, Green and Longman).

Struck, A. (1902), 'Zur Geschichte der Eisenbahnen Griechenlands', *Deutsche Rundschau für Geographie und Statistik* 24, 193–204.

Τζετζος, A. (1885), Σχεδιον της Πολεως των Πατρων 1:3000 (Πατρας).

Wagstaff, J. M. and Frangakis-Syrett, E. (1992), 'The port of Patras in the second Ottoman period: economy, demography and settlements, c. 1700–1830', *Revue du Monde Musulman et de la Méditerranée* 66, 79–94.

Χουλιαρακη, Μ. (1974), Γεωγραφικης, Διοικητικης και Πληθυσμιακης Εξελιξεως της Ελλαδος, 1821–1971, Τ. Α, μερος II (Αθηναι: Εθνικοι Κεντροι Κοινωνικων Ερευιων).

Υπογραφειου Οικουομιχων. Γραφειο Στατιστικης (1894), Εμποριον της Ελλαδος ... (Αθηναι).

9 Cityport Development and Cultural Heritage: The Case of Thessaloniki, Greece

STELLA KOSTOPOULOU

Department of Economics, Aristotelian University of Thessaloniki, Greece

Towns are largely dependent on their geographical position which deeply affects their relations with the outside world. Among the strongest factors affecting urban location and growth is transport, and transport nodes were considered highly significant by early location theorists (Cooley, 1894; Weber, 1899). Ports, as places where each-way exchanges between land and water transport regularly take place, represent one of the earliest forms of transport node. Because of their advantages for trade, cityports became financial and industrial centres, where transport and trade exert their influence on the social, cultural and economic development of the regions in which they are located.

Historically, port development has come about largely as a result of progress in and the needs of maritime transport. This relationship changed in the late 19th and through into the 20th century, in the context of the changing nature of traditional port-area industries; the role of maritime commerce, which became less critical and fundamental; and urban concentrations in coastal zones, which led to a diverse and sophisticated range of land and water uses taking place in an intrinsically delicate natural environment. Now, as cities find elements of their inherited infrastructure decaying and an explosion of suburban growth choking off the possibility of expansion at the periphery, planners find the waterfront increasingly attractive as a site for development. Ports are today, to a much greater degree than in the past, characterised by a diverse set of multiple, and often competing, social uses. This current round of revitalisation of urban port areas is the latest in a series of cycles of development which emerged over time as urban functions, available technology and social forces changed.

As a result, port-area evolution is characterised by growing complexity. It is necessary, however, to make a distinction between different types of water-oriented cities. In the redevelopment of port areas, different regional contexts can be identified all over the world, since the significance of water has varied between

Cityports, Coastal Zones and Regional Change. Edited by Brian Hoyle.
© 1996 John Wiley & Sons Ltd.

different urban cultures. In this context, the Mediterranean cityport differs because of its strong historic character and cultural heritage which does not readily admit 'standard' revitalisation projects. Thessaloniki, because of its long history, may be considered as a typical example of an historic Mediterranean cityport, today requiring specific planning attention. This chapter explores the history of the development of the cityport and the need for fresh new planning ideas, within the general framework of the evolution of cityports and current trends in coastal zone management.

THE HISTORICAL DEVELOPMENT OF WATERFRONTS

The fascinating history of the growth of port cities reaches back to ancient times, when ports were solidly built on a grand scale. Because of the availability of cheap and unlimited manual labour, docks, sea-walls, breakwaters and other harbour works received great attention (Quinn, 1972). From ancient times until today, the waterfront has been a focus for human habitation, going through many stages and major changes (Hoyle, 1988; Vallega, 1992).

The primitive cityport or mercantile stage

This was a very long stage that survived up to the early 19th century, until the first Industrial Revolution took off. During the mercantile stage, which developed from the 16th century in a European context as world trade grew because of the routes opened up by oceanic exploration, the most advanced cityports benefited from an external environment much more extensive than in earlier eras (Vallega, 1992).

The expanding cityport or palaeo-industrial stage

At this stage, with the technological changes of the first Industrial Revolution, the volume of shipping grew, the demand for more vessels became apparent and expanded port facilities became necessary, exerting a marked influence upon patterns of urban land use. Because of industrial development, many ports were forced to break out of their traditional confines, and the seeds of port–city separation were widely scattered (Hoyle, 1988).

The modern industrial cityport or neo-industrial stage

This stage, which took off in the late 19th century, was marked by the introduction of major technological changes in transportation, including eventually containerisation, which required extensive land sites, thus accelerating the separation of port–urban land uses and functions (Hoyle, 1988). Industrial growth reinforced its influence because of the need to establish industrial activity inside and near port areas. The major ports became workplaces articulated into the city, but held at a distance from its centre, creating their own territory. As greater and greater areas were swallowed up by port development, their morphology became chaotic and a source of pollution (Rebois, 1993).

Retreat from the waterfront or post-industrial stage

During the last decades of the 20th century a world-wide deconstruction of industrial cities has influenced their entire functioning and socio-economic equilibrium and has showed its greatest territorial extent in port cities, rendering obsolete, one after the other, buildings and facilities which were the wealth and pride of former times. New port terminals and manufacturing areas were located far from the old port areas and general cargo and industrial facilities were moved from the old areas to new ones (Hoyle, 1988). The separation between the port and the city thus reached its peak, causing widespread social and spatial effects and patterns. Within 20 years, port industrial zones, which were once hives of business, movement and production, were transformed into zones of urban decay.

Redevelopment of the waterfront

The rapid evolution of the technology of port activities, and the progress of social issues with more attention paid to environmental problems, have together encouraged cityports to re-utilise abandoned areas and revitalise unstable local economies. With a general scarcity of development sites, city planners and developers have seen the redevelopment potential of older, waterfront buildings and sites which were under-used or which contributed to air and water pollution (Chaline, 1993). New kinds of non-port and non-industrial functions were located in the waterfront area, leading to a wide spectrum of spatial re-organisation, new waterfront landscapes (Vallega, 1992) and the rise of new urban zones on the sites of decaying piers.

Nowadays, waterfronts on every continent are coming alive with new uses since every port city in the Western world seems to be infected by an epidemic of waterfront redevelopment. Waterfront projects form their own ecologies, overpowering the connection between the natural systems on the land behind and those on the water's edge (Bender, 1993). The scale of these developments, the magnitude of the social changes, and the past century's extraordinary impact on the environment call for fresh attitudes to port-area planning.

As a consequence of these phenomena there has been in recent years a discernible shift in the general principles and approaches used by planners and decision-makers in managing coastal environments especially in urban areas. The concept of *port and harbour management* has been replaced, to a great extent, by principles directed more fundamentally at *urban coastal zone management and estuarine conservation* where public demand for access to amenities and heightened concern for the environment are the newest forces for change, promising to shape the waterfronts of the 21st century (Bender, 1993).

COASTAL ZONE MANAGEMENT: THE NEW ERA

Because waterfront redevelopment was born as a project-oriented phenomenon, the literature did not initially take into account the planning dimensions within the wider context of the coastal area, including the environmental implications and the role of the cultural heritage (Vallega, 1992).

Planning dimensions

The traditional view of port evolution was that of ports as human settlements evolving through ports as transportation points, to ports as global goods distribution centres. This view, however, was limited in that ports were seen as structures in isolation, and not in the broader environmental context, as components of a larger ecosystem which is essentially subject to man's activities (Vandermeulen, 1992). This new identification of the urban coastal zone as a highly dynamic multifaceted zone of interaction between land and sea, overlain by new anthropogenic pressures, has redirected the dominant management approach in world-wide coastal environments towards that described as 'area-wide management' (Bowen, 1992), which involves broad coastal areas and uses ecosystemic principles.

, However, this ecosystemic approach, together with more general public and political perception of the need for greater coastal environment protection, may impose increasing constraints on the ability of commercial ports to expand. Thus, the challenge to port planners the world over is the successful integration of environmental pressures, both on the port and by the port, with all the problems of port development within the ever-changing context of international trade. Port-area planning is required to investigate both present and expected relationships between facilities, with the aim of realising which economic organisation is capable of producing the highest productivity levels and the greatest attractive power (Vallega, 1992). Bender (1993, 33) has suggested two criteria in waterfront planning: one addresses the scale and complexity of the parts, that is projects accommodating a wide range of uses and users that adapt their form to their place; the second suggests an approach to infrastructure and the natural systems of places, where future planners should recognise topography and native ecologies as opportunities rather than problems.

Environmental problems

In the ports of the 20th century, increasing environmental problems such as chemical and sewage pollution, disruption of fisheries and the storage of hazardous materials have gained growing public awareness. Impacts of ports on their environment are probably most easily visualised as being of three kinds (Vandermeulen, 1992): they act as concentration points for pollutants, they compete with other existing ecosystems, and they affect adjacent human coastal communities. The environmental costs of these pressures are often difficult to translate into economic costs, one reason being that ecological and economic values differ philosophically and use different currencies. There is a need for the two principal dimensions, technological and environmental, to be linked philosophically and economically and for a standardisation of national and international quality criteria (Vandermeulen, 1992).

The role of the cultural heritage

By its nature, the waterfront is commonly the area in which some of the most important elements of a cityport's history can be found. In recent years, with the implementation of a wealth of projects and initiatives, the need for the urban redevelopment of historic ports has found favour in locations where urban centres of major historic importance and ports sit side by side. However, historic waterfront redevelopment prospects differ, according to countries and regions and to the role that the cultural heritage plays in the establishment of development strategies and rejuvenation plans (Vallega, 1992).

For example, in Japan, the need to protect an historical legacy does not exist, because cityports have changed their waterfront several times in recent decades giving new shape to older spatial patterns. In the United States, despite the strong growth processes, waterfront plans are sometimes seriously conditioned by the historical remains of earlier cityports. In Northern Europe, where to a greater or lesser extent waterfront redevelopment has been due to the decline in cityport activities, the question is whether and to what extent older settlements and port structures may be employed as a tool for creating a new attractive waterfront. In the Mediterranean coastal zones waterfront redevelopment involves a limited number of cityports, proceeding more slowly than in Northern Europe and along the Atlantic coasts.

In the Mediterranean, described by Braudel (1979) in Eurocentric terms as 'the initial focus of the world', many continental and island cityports are still characterised by areas of archaeological value from Greek and Roman times. In most ancient Greek and Roman cities the market (*agora*) was situated at or near the port, which was not only the main centre of economic activity but also the social, cultural and political focus of urban life. This tradition was reinforced in the 14th and 15th centuries, when the Mediterranean acquired major importance as an urban-based theatre of international trade. Mediterranean port cities such as Venice, Genoa, Naples, Ancona, Trieste, Marseille and Barcelona produced a special kind of port area where public city-life took place, and was combined with a special urban form and architecture. With the development of steamships and the new political, economic, military and technological conditions, the fortified port-cities disappeared and new urban areas opened towards the sea, towards their hinterlands and towards international flows of people, goods and ideas (Meyer, 1993).

Vallega (1992, 57) has proposed that such cases can be classified into a specific version of the Mediterranean mode, the archaeology-based Mediterranean waterfront, where this extraordinary cultural heritage should be regarded as a basis for planning specifically concerned with this unusual cultural context, and where the essential objective is to redevelop but not to rejuvenate. Thessaloniki, due to its long history going back to ancient times, may be characterised as a typical example of such a history-based waterfront which needs specific redevelopment attention.

·THE CASE OF THE CITYPORT OF THESSALONIKI

Thessaloniki, the focal point of Macedonia (Greece), lying on the slopes of the south-western extension of the Balkan massif, is located on the inner recess of Thermaikos Gulf, an inlet of the Aegean Sea. The city occupies a semicircle and is bordered by the deltas of the (no longer navigable) rivers Axios and Gallikos to the west, a 600-foot Acropolis to the north and rolling hills to the east. Facing the amphitheatre of Thessaloniki from the south, in the middle of the urban shoreline, is the deep, spacious and protected harbour (Figure 9.1) (Baxevanis, 1963).

The port of Thessaloniki, primarily looking to a continental hinterland, is the terminal of a system of land communications and lies on the shortest route between central Europe and the Middle East (Hammond, 1972). Due to its strategic location in the most favourable geographical position in Macedonia, the gateway to the Balkans, Thessaloniki from its foundation until today has never lost its early importance as a large commercial and transit centre, exerting a major influence throughout the eastern basin of the Mediterranean.

Today, Thessaloniki is the second largest urban area, with a population of nearly 970 000 (1991), and the second largest concentration of industrial activity in modern Greece, with manufacturing a decisive parameter in urban development. At the regional level, Thessaloniki is the industrial, manufacturing, financial, banking and general economic capital of Northern Greece, connected with the main economic centres of Eastern and Western Europe by significant road, rail, air and sea networks. The port of Thessaloniki is the leading exporting and transit port of Greece, accounting for more than 50% of all Greek exports, and the second regarding total traffic. Some 98% of the port area operates under a Free Zone status, serving Greek imports–exports and also, to a significant degree, the transit trade of neighbouring Balkan countries, a role enhanced in the 1990s by the disintegration of former Yugoslavia.

THE HISTORIC EVOLUTION OF THE CITY AND PORT OF THESSALONIKI

Ancient period

During the long history of Thessaloniki, the port has been closely related to the economic and cultural life not only of the city and the surrounding region, but

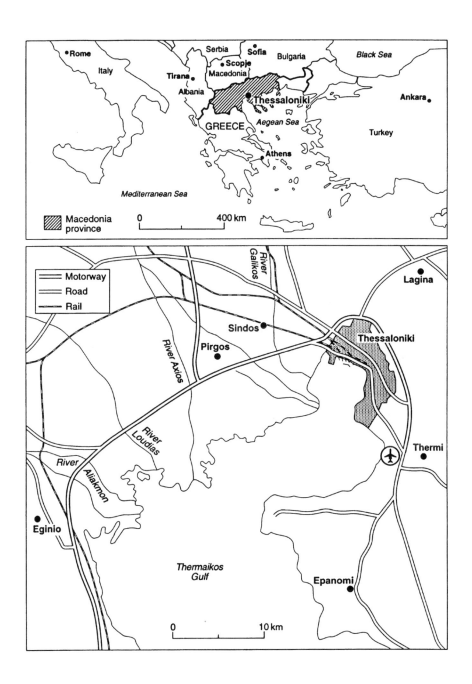

Figure 9.1. The site of Thessaloniki, Greece

also of the broader Balkan area. The history of the city runs parallel to and almost coincides with the life of its port. Thessaloniki was founded at the beginning of a great era in the history of ancient Greece and, in more general terms, of mankind, the so-called Hellenistic Age. In that epoch, following the conquests made by Philip II and Alexander the Great, Greek civilisation spread to much of the then known world. Macedonia, the land of its kings, needed close contact with these far-distant lands, and with an extensive and wealthy hinterland which was looking for a natural outlet to the sea. King Kassandros, appointed by Alexander the Great as the caretaker of the Macedonian state during his absence, pointed out the innermost deep recess of Thermaikos Gulf which had many natural roadsteads (Vacalopoulos, 1963). Thus, Thessaloniki was founded in 316 BC by Kassandros who named the city after his wife, half-sister of Alexander the Great. The walled city of Thessaloniki in Hellenistic times comprised the area around and above its harbour. As the port of the Macedonian state capital of Pella, Thessaloniki was the base for her mercantile marine and navy, accommodating the substantial Macedonian fleets. The commercial development of the town was rapid and substantial, and the city soon gained the reputation of being the 'Mother of all Macedonia', a commercial centre with close trading connections with all the ports of the East (Moutsopoulos, 1980).

Roman period

After the occupation of Macedonia by the Romans, in the mid-2nd century BC, Thessaloniki became the capital of the Roman *Provincia Macedoniae* (Papahadjis, 1957). During the Roman era, Thessaloniki was a Free City (*Civitas Libera*), preserving the Greek language and its ethnic integrity, developing into the most populous city in Macedonia – its metropolis, according to Strabo. The growth of Thessaloniki as a commercial port continued when the Via Egnatia, the shortest route from the Adriatic to Constantinople, was routed through the city, which soon became the most important military and commercial station on this great imperial road and its port one of the most important of the entire Roman Empire (Vacalopoulos, 1963).

From existing historical evidence, it appears that the port of Thessaloniki was systematically developed at the end of the 4th century during the reign of Constantine the Great. According to the historian Zosimus, the port and city were founded by King Kassandros of Macedonia about 315 BC who named the new city port after his wife Thessaloniki, sister of Alexander the Great. No port had previously existed on this site, but the new facilities were soon able to accommodate 200 warships and 2000 freighters (Letsas, 1961). An artificial harbour was constructed at the south-western end of the city in 324 AD, including dockyards to protect loading and unloading facilities against the south and south-west winds. The artificial port, then lying outside the walls of the town, consisted of a wide closed basin and, protected by a breakwater, it formed a square with its entrance on the east (Figure 9.2).

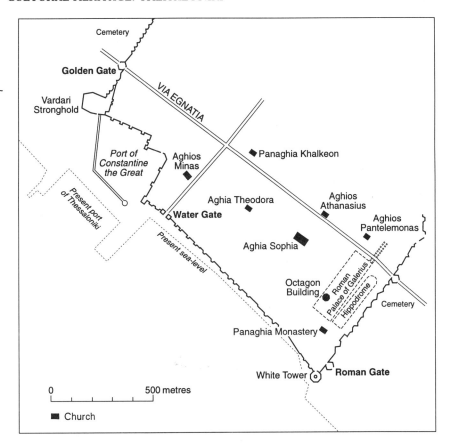

Figure 9.2. The port of Thessaloniki in the Roman and Byzantine periods (4th–5th centuries).

Byzantine period

During the Byzantine era, Thessaloniki became in population terms the second largest city of the Byzantine Empire, and remained the most important commercial, administrative and cultural centre in the whole of the Balkans. It was the busiest port in the Balkan peninsula, where Slav peoples came to exchange their produce for goods which reached Thessaloniki from the remotest places of the Far East (Baxevanis, 1963).

Because of its location and importance, the cityport experienced the destiny of all the great harbours of the Mediterranean which were frequently the objects of attacks by pirates and corsairs. During the whole of this period, the city suffered from the invasions of Arabs, Saracens, Normans and Slavs, and trade with the hinterland was often interrupted. In spite of these calamities, the port of Thessaloniki showed continuous development and soon the town regained its earlier splendour, owing to its crucial geographical position and to the vigour and efficiency of its inhabitants. From the 12th century, Thessaloniki's position as an important port of entry and departure of goods to and from the hinterland was further enhanced by the settlement of European traders, including Venetians and Genoese (Hoffman, 1968).

In the 13th and 14th centuries the decaying and disintegrating Byzantine Empire was unable to control much of the hinterland and, as a result, Thessaloniki's ties with the hinterland weakened and trade often came to a standstill. Finally, Thessaloniki fell into the hands of the enormous army of Murad II on 29 March 1430 and remained under Turkish rule until 1912 when, after the successes of the Greek army over the Turks during the Balkan wars, the city became Greek again.

Turkish period

During the first decades of the Turkish occupation, nearly all commercial activities declined and Thessaloniki was isolated from both Western and intra-Balkan commerce. With time, however, the city started to enjoy the fruits of a politically united hinterland and economic and cultural freedom from the Moslem. Its trade, manipulated by shrewd and energetic merchants, was large and rivalled that of Constantinople, the capital of the Ottoman Empire. In the 15th century, a Jewish community was established and developed at Thessaloniki, reinforcing the commerce of the city in important ways.

In the 16th century the port experienced some re-orientation, like all other ports in the Eastern Mediterranean, due to the discovery of the direct sea-route between Europe and the Far East and the discovery of America, which re-directed international trading interests from the East to the West. During the following century, though, interest was once again focused on the East, due in part to the signs of decline of the Ottoman Empire. In 1700 new harbour works were constructed and foreign shipping agencies were established in the town, representing most of the Western European countries. At the end of the 17th century and during the whole of the 18th century, Thessaloniki experienced a remarkable commercial traffic growth as the port served the trade of the Ottoman Empire, developing into the primary transit port of the Balkan peninsula. Figures from Italian and French records show that on average over 1000 ocean vessels a year entered Thessaloniki between the years 1738 and 1848 (Svoronos, 1956).

The port was now busier than ever and was considered quite safe, not only because it was situated in the innermost part of the deep bay dominated by the city, but also because it was protected by the wall which ran along by the sea and by the wall's huge towers on its western and eastern extremities, equipped with enormous guns. Space from the artificial Roman/Byzantine port, which was filled in during the first years of the Turkish occupation, comprised the only part of the city outside the walls (Figure 9.2). Port facilities, next to the Quay Gate which connected them with the commercial centre of the city, consisted of a small wooden wharf and associated installations including wholesale stores, retail shops and workshops related to the port traffic.

By the end of the 19th century, the Great Powers which had commercially penetrated the Ottoman Empire thought that a railway network was necessary in order to facilitate trade with Central Europe. The railway station was built in 1871 at the western end of the port, next to the wholesale and retail market and the business and industrial areas. Thessaloniki was linked by rail with Skopje in 1871 and Belgrade in 1880, thus reaching the main European railway network; and with Monastir in 1893 and Istanbul in 1895, thus completing one of the Trans-Orient Express routes (Papagiannopoulos, 1982). As the railways rapidly criss-crossed the hinterland, traffic became heavier than ever, and new channels of distribution were established.

The re-organisation of the port area began in the 1870s, when because of the growth of Macedonian trade especially with international markets, and because of pressure from the city's trading houses, the Turkish authorities proceeded to demolish the ancient sea-front wall and thus the city was opened to the sea. The new waterfront was paved with cobblestones and provided with a straight stone breakwater, new trade storehouses and a custom-house (1875) (Figure 9.3). The demolition of the sea-wall and the construction of the waterfront is the starting point of a new era for the city, defined by her renewed local and international role (Vougias et al., 1993).

In the last two decades of the 19th century, the port of Thessaloniki, as the main gateway to the Balkans and Central Europe, was handling 14% of the volume of trade in the Ottoman Empire. In 1881, 345 steamships and 5318 sailing ships visited the port (Papagiannopoulos, 1982). The merchant fleet belonging to the occupied Greeks had held first place in the port's traffic since the end of the 18th century. When steamships appeared in the Mediterranean, the Greek share fell to just 1% of the goods transported, leaving almost all trade to English, French, Italian and Russian steamships. Thus the Greek domination of the sea, based on small, slow sailing ships, ended.

By the end of the 19th century the port started to assume its present form. In 1888, the construction of a modern artificial port was proposed to provide new berths and to protect shipping from bad weather conditions. In 1896 the licence for the construction and exploitation of the port was given to the French Port Company, which retained the franchise of the port even after the eventual liberation of the city from Turkish rule. These port works started in 1899, when electricity was introduced into the town, and were completed in 1901, when the new harbour was opened for navigation. In 1904 further harbour works were undertaken and new railway branch lines and grain silos facilitated cargo handling at the port. Port facilities were further enlarged, and the Customs House, warehouses and offices of large marine and commercial companies, the French Port Company and banks were installed in the port area (Figure 9.3). In 1907 the first electric trams started operating and in 1916 the city was connected to the Greek railway network which extended as far as Athens.

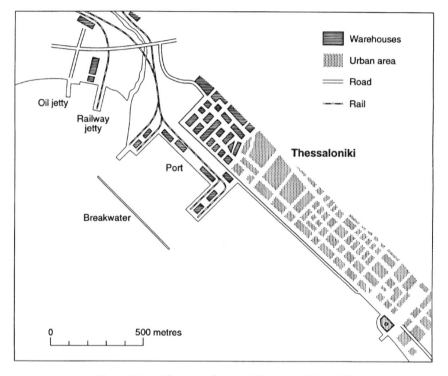

Figure 9.3.   The waterfront at Thessaloniki in 1909

Contemporary period

Between 1878 and 1920 the politico-geographical map of the broader Balkan area
was completely redrawn. The liberation of the city from the Turks in 1912, the
end of the Second Balkan War, the expansion of some existing states (Serbia,
Bulgaria) and the creation of Albania, all changed the situation radically. The
changed geopolitical situation decreased the use of the port of Thessaloniki
considerably, because of the new states' nationalistic policies and the
development of their own harbours (Vasdravelis, 1960). The port of Thessaloniki
experienced a steep decline and became confined to trade within the Greek
borders only. Nevertheless, the city remained the principal natural entry port for
a large region. To combat the impact of these territorial changes in the
hinterland, the Greek Government established in 1914 at the port of Thessaloniki
a Free Zone, which started to operate in 1925, granting many privileges in order
to facilitate trade in the Balkans and the development of manufacturing industry.
Moreover, because of an old promise to Yugoslavia and in order to maintain the
long-established friendship between the two countries, an area of 94 000 m^2 was
ceded in 1929 to a Yugoslavian Free Zone with the sole purpose of serving
Yugoslavian trade. This arrangement continued until 1975.

During this period, two major events changed dramatically the built and social environment of the city: the fire of 1917 and the arrival of refugees in 1922 after the calamity in Asia Minor, both of which increased the commercial and other urban needs (Karadimou-Yerolimpos, 1975). The fire of 1917 destroyed nearly the whole of the city centre, including the busy commercial sector. However, the port, for the salvation of which a great effort was made, was left untouched. The city plan which was worked out afterwards, the famous Herbrad plan (1923), represented a radical intervention in the city's historical evolution process, even though the city's ancient grid system was kept in its basic essentials. The central section retained its commercial, administrative and cultural roles, not to mention its role as the site of the city's most imposing buildings. Thanks to the magnificent promenade created in the port area, residents were able to enjoy the sea air to the full, an advantage not shared by all coastal cities. A 'piazetta' was planned at the western end of the sea front, and was intended as a place of refreshment and relaxation. Unfortunately, Herbrad's original plans were not carried out, for political reasons, but the city owes its present form to the then Liberal government's adoption of the Herbrad plan in principle.

Until the implementation of the 1921 city plan, the port with its quays (Figure 9.3) provided the only large open collective area in the city, a place of public urban, market and leisure activities of the sea-front zone. With the implementation of the plan, the port was pushed westwards with its docks and warehouses grouped around it and thus the port area was cut off from the surrounding urban grid and the nearby wholesale market, consisting of a sort of barrier to the sea for the western part of the city. In 1930 the assets of the French Port Company were bought off and the Port of Thessaloniki Trust Fund was established for the financing of new port works. In 1933 a plan for the extension of the port to the west and the construction of the sea-front avenue to the east was prepared (Thessaloniki Port Authorities, 1963).

At the end of the Second World War the Germans, before leaving, blew up every harbour installation and establishment. The port was destroyed to such an extent that all old and new works, in the harbour and on land, lay in ruins. A period of restoration, extension and modernisation of the port area immediately started. In a series of consecutive projects, the port area has been tripled and the port has been equipped with the most modern mechanical equipment (Figure 9.4). Modern buildings housing the offices of the Customs and the Port Authority and special buildings housing the workshops were built. The harbour basins were cleared of shipwrecks and a radio-direction system for the safe navigation of ships was installed. In 1953 the Free Zone and the Port Trust Fund merged under the name of Free Zone and Port of Thessaloniki aiming at the coordination of the activities of its two predecessors.

As far as port-oriented activities are concerned, there were at this time few firms whose business was derived solely from the port and hence few were located near the port or on the waterfront to take advantage of waste disposal, lower transfer costs and proximity to cheap level land. Purely port-connected

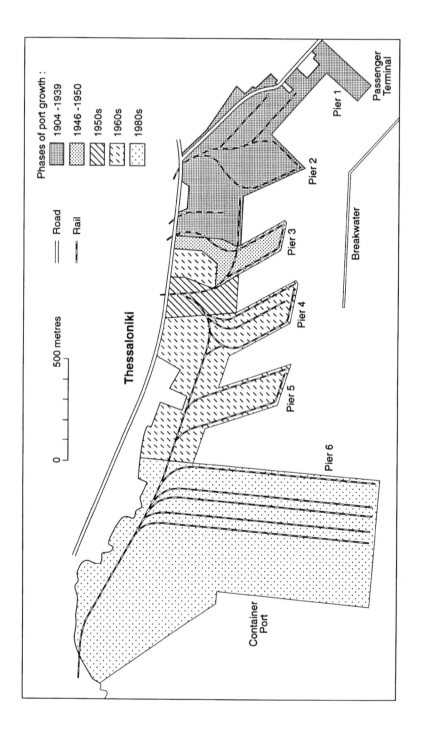

Figure 9.4. The modern port of Thessaloniki

manufacturing was found exclusively in the Greek Zone; this took advantage of customs-free imports of raw materials and did not compete with the domestic industries supplying the home market. Except for a large brewery, refrigeration plants, slaughterhouses and tanneries, no other industries were located near the waterfront (Baxevanis, 1963).

In 1970, the Organisation of the Port of Thessaloniki was founded, a self-governing public organisation, with exclusive rights to run the port under the supervision of the Ministry of the Merchant Marine. Thessaloniki Port Authority has always been financially self-sufficient, utilising considerable sums from its surplus funds, in addition to European Union assistance programmes, for extension projects and the improvement of technological equipment. A container terminal was recently constructed, designed with the latest technical standards, and thus the port now covers all modes of cargo handling (Figures 9.4 and 9.5). In addition, the Port Authority has plans to purchase new mechanical equipment that will significantly increase its potential. On completion of these works, the port of Thessaloniki will further develop and is expected to contribute to the expansion of the city's economic sphere of influence. With the broader geopolitical and economic changes in Eastern Europe, the port has expanded its trading activities and is expected to respond positively to the growing demand (Thessaloniki Port Authorities, 1992). A new element during recent years is the dynamic growth of passenger traffic with the Aegean islands, which is increasing spectacularly, elevating Thessaloniki as the leading passenger port of Northern Greece. The startling rise in the number of passengers (88 000 in 1989; 200 000 in 1991) caused the transformation of a part of the Customs House into a passenger station.

Apart from its purely economic contribution to the life of the city, the port organisation is actively involved in cultural and other events. An event that changed the traditional perception of the port area was the partial and temporary use of some buildings for cultural events during the 'Biennale' in 1986. These activities have continued and expanded since then, and have gained a more permanent character, giving a new functional importance to the port and involving the removal of the port commercial activities to the new west pier and the modernisation of the installations.

The recent changes in the technology of transport, together with the expansion of the sixth pier (which doubled the total land surface of the port), the modernisation of the loading/unloading and storing installations, the construction of new road and rail links, the ever-increasing demand for new passenger sea-links and the new cultural uses of the built equipment, within the general framework of the international geopolitical evolution, imposes the necessity for a restructuring of port installations and a general rearrangement of the organisation of the port area, so as to fulfil the new demands.

THE REDEVELOPMENT OF THESSALONIKI'S WATERFRONT

During the last 30 years, a series of projects for the rejuvenation of the urban

Figure 9.5. The port and city of Thessaloniki. (Courtesy Port of Thessaloniki)

coastal zone have been proposed which, however, have not been realised, except for the extension of the waterfront to the east which started in 1960, financed by the Port Fund which gained most of the profits derived from the filling in along the new sea avenue (Vougias et al., 1993). The work was limited to the construction of a rather dreary straight paved breakwater and a park. In 1966 a proposal for the infilling of the old waterfront caused some reactions from the Municipality, the National Organisation for Tourism and other public organisations, who pointed out the need for a general urban plan including the seafront of the city. The idea led to a proclamation in 1971 of a national competition for the development of the coastal zone along 8 km. None of the proposals received was accepted in its entirety, but in 1976 the Municipality decided to prepare its own proposal, using selective elements from the proposals submitted to the 1971 competition: the filling in of the old 150-m wide waterfront and the construction of an underground car parking area with 3000 places. In 1978 an international competition for the planning and construction of the work was proclaimed. Only VTN Los Angeles Multinational submitted, proposing the reduction of parking places to 1870 and the enlargement of the urban coastal marina, and the competition ended without result. During the next years interest was focused on more wide-ranging planning regulations for the whole city. Within this context an underwater highway was proposed at some distance from the shoreline, but the idea of extensive parking facilities was rejected.

The proposal for the filling in of the old waterfront was put forward again in 1986 by the Municipal Authorities, who characterised it as a major urban development plan. In 1988 a national competition was proclaimed for the infilling with an underwater highway and 6000 parking places. Because of the strong reactions of administrative and scientific organisations (Technical Chamber, universities, architects confederations, etc.) the competition was once again unproductive. In 1990 another proposal was made from the very same municipal authorities and the Ministry of the Environment, in some respects similar to previous ones. This proposal, focusing again on the construction of an underwater highway and a large parking area on the infilled part of the old waterfront, was mainly an urban traffic and parking plan, underestimating the importance of the historical, architectural and environmental aspects.

The need for the construction of an underwater highway was a common argument for all relevant civic public bodies and it was included in the 1985 Urban Plan of the city which did not however refer to the construction of an underground parking area or infilling of the old waterfront. On the contrary, the Urban Plan's main perception was the preservation and elevation of the historical character of the city centre. The necessity of the underground coastal highway was also based on the preliminary results of a General Urban Traffic Study, that the construction of such an artery could, theoretically, reduce the city-centre traffic by approximately 50%.

According to the proposals of the 'Traffic Study for the Expansion of the Old Waterfront' that evaluated the origin/destination data of the General Urban

Traffic Study, the underwater highway should be at a distance of 80–100 m from the old waterfront, with a width of 25 m, and the underwater garage should have three two-floor parking zones of 800, 1700 and 1600 places respectively. The study also proposed the creation of 'archaeological paths' with the primary one along the old waterfront, that is exactly the part to be destroyed by the proposed work. Thus, the study concluded with the strong contradiction that the infilling of the old waterfront was to be included within one of the basic archaeological paths of the city. The study caused very strong reactions and has been a subject of debate for several years, fortunately without any development along these lines being realised.

· No further major interventions have been made until recently when Thessaloniki was nominated as the 1997 Cultural Capital of Europe. Within this concept, a plan was set up for the renovation and re-orientation of the port area as a cultural centre (Vasiliadis et al., 1994). The plan suggests the incorporation of new cultural functions with the overall traditional activities of the port, and the closer relation of the port area to the general urban settlement (transport networks, historical character of the city centre, development of zones surrounding the port, etc.).

· In the 'Thessaloniki Cultural Capital' plan for the renovation of the port area, the general aims are: the management of the port area as a uniform zone functionally, historically and in urban planning terms; the restoration of the organisation of historical space in the older part of the port and the conservation of its particular morphology; avoidance of future demolition of old buildings and the incorporation of new buildings within the overall existing built environment. Consequently, new land uses are to be incorporated within the existing port activities in such a way that port (port administration, passenger traffic) and urban (public and cultural) uses can coexist.

The new functional groupings are to include a Congress Centre, a Centre of Music, a Cinema Centre and an Exhibition Centre (Figure 9.6). Each functional entity will consist of a central building, which could remain as a permanent transformation, and some support buildings, which could regain their previous uses after the end of the 1997 'Thessaloniki Cultural Capital' year. Thus, three types of area uses are proposed: (a) temporary, for the one-year period of 'Thessaloniki Cultural Capital'; (b) transitory, shortly before and after this one-year period; (c) permanent, after the Cultural Capital year, which demands a more general plan. These new planned activities are expected to create traffic implications to be confronted within the general framework of urban plans. It is estimated though that no major parking problems will be caused, due to the existence of parking areas in the surrounding zone, and also to the large free land areas within the port zone.

· As such the '1997 Thessaloniki Cultural Capital of Europe' phenomenon could act as an important catalyst in the process of port-area renovation, comparable with the 1992 Columbus Exhibition at Genoa. In the case of Genoa, within the general perception of the redevelopment of the historic port as a key component

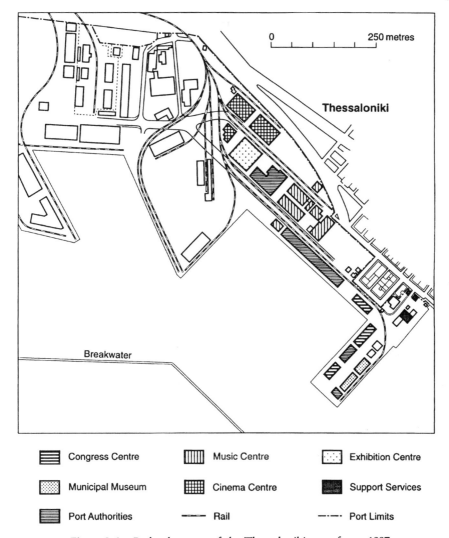

☰ Congress Centre	▥ Music Centre	░ Exhibition Centre
▦ Municipal Museum	▦ Cinema Centre	▓ Support Services
▤ Port Authorities	—— Rail	—··— Port Limits

Figure 9.6. Redevelopment of the Thessaloniki waterfront, 1997

of the city's historic centre, the necessity to define alternative uses for the urban port spaces was stressed without, however, running to the opposite extreme: that is, a city which invades the port with residual urban functions totally alien to the nature of the waterfront, and to its pre-existing historical sites (Capocaccia, 1993, 85). Thessaloniki's plan for the redevelopment of the waterfront seems to have the same general perception, to maintain the traditional range of functions, adding new functions to the traditional ones, i.e. to create a waterfront which will offer new opportunities of accessibility and use to both city residents and commercial agents.

CONCLUSIONS

Thessaloniki's port thus remains a place with an historically continuous function, with intensive trade activity and without any permanent downgrading of the installations. The particular identity of Thessaloniki derives from its unique characteristics which survived through the successive transformations of the urban grid, where the relationship between the port and the city is undoubtedly one of the major components of the urban landscape. Thessaloniki was and is not a nautical city, with no major shipbuilding or repairing activities, and her inhabitants were not primarily sailors, skippers or shipowners. Their relation with the sea was one of use and experience, of trade and transhipment (Vougias et al., 1993).

The port area has retained its importance for the city and for most of the older topological zones such as the market, the railway station and the industrial area. Since its establishment, the waterfront has functioned as a place of extension of commercial and storage activities. Thessaloniki traditionally represents the typical Mediterranean port where the market was related to the port transhipment function and infrastructure, and which provided the focus, the entrance and exit point of the city, the place of movement of people, goods and ideas and contact with unknown faraway places. The waterfront, together with the broad quays, provided the only collective open space of the city, the ideal place to stimulate and cultivate the spirit of freedom, of openness, of open-minded interest in strange cultures. The demolition of the sea-wall and the construction of the port introduced new building and architectural types in the image of the city, directly related to the ideas of modernisation and development. Since the end of the 19th century and the first decades of the 20th, this area gained the reputation of the cosmopolitan centre of the city, with economic, political and cultural functions. Following the damage from the Second World War, the port area comprises most of the historical centre of the city that was saved from the 1917 fire, a fact that imposes the necessity for its preservation and protection.

⋅ During recent years the port has been established as a functional point of contact between the eastern and western parts of the city. It occupies the most easterly part of the historical centre with which it is related through the dense network of trade and transhipment services, located to the north and east of the port. It includes the terminus of the long urban coastal zone, through which the contact and the outlet of the eastern part of the city to the sea is assured. West and north of the port, there are extensive industrial zones and major road and rail links.

⋅ It must be underlined that because of the above functional relations between port installations and the basic activities of the city, serious transport problems are caused, due to the railway lines going through the surrounding area and the large number of vehicles serving the port, plus the environmental problems such as the danger from the movement of hazardous or polluting cargoes. Nevertheless, these problems do not alter the functional relationships of the port with the city, which can be characterised as stable during their long history. These relationships impose the necessity for any redevelopment plan to retain the historical incorporation of the port within the urban environment, to emphasise the waterfront as an area of public domain, to re-emphasise the role of the waterfront as a citadel of the city's civilisation, to consider the historical heritage as the basis for the port-area planning and to regard the waterfront as a leading component of the urban coastal zone.

REFERENCES

Baxevanis, J. (1963) *The port of Thessaloniki* (Thessaloniki: Institute for Balkan Studies).

Bender, R. (1993), 'Where the city meets the shore', in Bruttomesso, R. (ed.), *Waterfronts: a new frontier for cities on water* (Venice: International Centre Cities on Water), 32–5.

Bowen, R.E. (1992), 'The role of emerging coastal management in port and harbour management', in Dolman, A.J. and van Ettinger, J. (eds), *Ports as nodal points in a global transport system* (Oxford: Pergamon Press), 229–37.

Braudel, F. (1979), *Civilisation matérielle, économie et capitalisme (XV–XVIII siècles), Le temps du monde* (Paris: Colin).

Capocaccia, F. (1993), 'Large urban projects for the port of Genoa: historic port and passenger terminal', in Bruttomesso, R. (ed.), *Waterfronts: a new frontier for cities on water* (Venice: International Centre Cities on Water), 84–7.

Chaline, C. (1993), 'Deconstruction and reconstruction in port cities', in Rebois, D. (ed.), *Building the city by the waterside* (Europan: Zaanstad Seminar) Tefchos, 11, 45–7.

Cooley, C.H. (1894), *The theory of transportation* (Baltimore: Publications of the American Economic Association, IX, No. 3).

Hammond, N.G.L. (1972), *A history of Macedonia, Volume 1: Historical geography and prehistory* (Oxford University Press).

Herbrad, D.R. (1923, 1927), 'Le nouveau plan de Salonique', *L'Architecture* 36(8).

Hoffman, W.G. (1968), 'Thessaloniki: the impact of a changing hinterland', *East European Quarterly* 2(1), 1–27.

Hoyle, B.S. (1988), 'Development dynamics at the port–city interface', in Hoyle, B.S., Pinder, D.A. and Hussain, M.S. (eds), *Revitalising the waterfront: international dimensions of dockland redevelopment* (London: Belhaven), 3–19.

Karadimou-Yerolympos, A. (1985), *The replanning of Thessaloniki after the fire of 1917: a keystone in the history of the city and the development of Greek city planning* (Published PhD Thesis, Thessaloniki, Municipality of Thessaloniki Publications).

Letsas, A. (1961), *History of Thessaloniki* (Thessaloniki).

Meyer, H. (1993), 'Port areas: cities meet the world', in Rebois, D. (ed.), *Building the city by the waterside* (Europan: Zaanstad Seminar) Tefchos, 11, 48–52.

Moutsopoulos, N.K. (1980), *Thessaloniki 1900–1917* (Thessaloniki: Molho Publications).

Papagiannopoulos, A. (1982), *History of Thessaloniki* (Thessaloniki: Rekos Publications).

Papahadjis, N. (1957), *Monuments of Thessaloniki* (Thessaloniki: Molho Publications).

Quinn, A.D. (1972), *Design and construction of ports and marine structures* (New York: McGraw-Hill).

Rebois, D. (1993), 'The water and the city', in Rebois, D. (ed.), *Building the city by the waterside* (Europan: Zaanstad Seminar) Tefchos, 11, 42–4.

Svoronos, N. (1956), *Le commerce de Salonique au XVIIIe siècle* (Paris).

Thessaloniki Port Authorities (1963), *The port of Thessaloniki* (Thessaloniki: Port of Thessaloniki Publications).

Thessaloniki Port Authorities, *Annual reports* (Thessaloniki: Port of Thessaloniki Publications).

Traganou, O. (1983), 'The beginning of industrial development in Thessaloniki', *Archaeology* 7, 97–102.

Vacalopoulos, A. (1963), *A history of Thessaloniki* (Thessaloniki: Institute for Balkan Studies).

Vallega, A. (1992), *The changing waterfront in coastal area management* (Milan: Franco Angeli: Ocean Change Publications).

Vandermeulen, J. H. (1992), 'Ports and environmental developments', in Dolman, A. J. and van Ettinger, J. (eds), *Ports as nodal points in a global transport system* (Oxford: Pergamon Press), 215–27.

Vasdravelis, J. (1960), *The Port of Thessaloniki* (Thessaloniki).

Vasiliadis, A., Papazoglou, A., Papamichos, N., Stavraka Chastaoglou, V. and Chatzimichalis, K. (1994), *The port of Thessaloniki: possibilities of housing development in the cultural capital of Europe 1997* (Thessaloniki: The Organisation of the Cultural Capital Publications).

Vougias, S., Kalogirou, N., Karadimou-Yerolympos, A., Papanikolaou, M., Patramanis, K., Chatzopouloa, G. and Kourakis, A. (1993), *The issue of the expansion of the old waterfront of Thessaloniki* (Thessaloniki: Technical Chamber of Greece Publications).

Weber, A. F. (1899), *The growth of cities in the nineteenth century: a study of statistics* (New York: Macmillan).

10 Cityports and Coastal Zones in Contemporary Africa: Mombasa and the Indian Ocean Façade of Kenya

BRIAN HOYLE

Department of Geography, University of Southampton, UK

Recent decades have witnessed substantial changes in port–city and cityport–region relationships. In the advanced world, and increasingly in developing countries too, the migration of port activities towards deeper water, as a consequence of technological change, has become an increasingly common phenomenon. This has introduced in many ports around the world an unaccustomed separation of port and urban functions, and the consequent redevelopment of older port zones and inner city areas (Hoyle *et al.*, 1988; Breen and Rigby, 1993). Similarly, the role of the cityport in a regional development context is beginning to receive increased attention, in terms of the impacts of rapid urban growth and port activity on neighbouring coastal environments, and in terms of the search for a more balanced, integrated approach to the management of port-city regions.

This chapter first presents a brief overview of the study of African cityports and their associated regions. The discussion then focuses upon the coastal zone of Kenya and the port of Mombasa, as an illustration of the interdependence in an African context of the cityport and its associated coastal zone, in terms of environmental contexts, historical legacies, political problems and planning issues. The inter-related objectives are, first, to review the characteristics of this case study in broad terms, and, second, to comment on the relevance of these trends to African cityports in general. These interdependent components lead to comments on the need for progress in the development of cityport–region theory, and in the analysis of a wider range of further examples from African and other developing countries, in order to identify promising avenues for future research.

Cityports, Coastal Zones and Regional Change. Edited by Brian Hoyle.
© 1996 John Wiley & Sons Ltd.

PORT CITIES AND CITYPORT REGIONS IN MODERN AFRICA

In Africa, the cityport has traditionally been perceived as a point of entry for colonial influences, and as a central place in terms of the interdependence of the metropolitan power and the overseas dependency, rather than in terms of the evolving economy of a dependent territory. In post-colonial Africa these perceptions have changed, and the peripheral location of many coastal cityport capitals has been highlighted in the context of independent Africa's new economic development structures and patterns (Mabogunje, 1989). Some African countries (Côte d'Ivoire, Nigeria, Tanzania) have attempted to alleviate the problem by relocating the political capital function more centrally within their national territory, well away from the coastal 'doorstep' location associated with the colonial past (Hoyle, 1979; Christopher, 1985; Kironde, 1993).

It is widely recognised, however, that port cities have played a large part in the development of modern Africa, notably through their role as nodes in international maritime transport systems. Although often characterised as gateway settlements from the standpoint of a colonising power intent on resource exploitation, coastal cityports also provided windows on a wider world for the societies and economies of coastal and interior Africa. The role of colonial port cities in African development has undoubtedly been highly significant; Arab and European cityport foundations in the 18th and 19th centuries, whether as innovations on virgin sites or based on older settlements established by earlier cultures, provided key elements both in transport networks and in urban systems. Subsequently modified during the process of 20th-century development, these nodes have often provided foundation stones of continuing importance in the context of modern urban and transport systems.

The study of African port cities, as of port cities in general, has traditionally been inclined to focus attention either upon the origins and evolution of port facilities, activities and trade structures, or upon the patterns and problems of urban growth, land use and functions. Most attention has been given to problems of urban planning, economic functions or growth problems in specific locations or regions, and the approach has tended somewhat introspectively to analyse the port-related and the urban aspects of port cities not only in isolation from one another but also somewhat separately from the socio-economic and political systems of the regions and countries to which they belong. The role of colonial port cities has been discussed at length (Broeze et al., 1986), as have maritime perspectives on African settlement and development (Stone, 1985); and the growth of individual port cities has been frequently and carefully recorded (Gillman, 1945; Schulze, 1970; Moorsom, 1984; Hoyle, 1985; Forjaz, 1992).

Attention has also been directed, however, to the characteristics of and policies associated with the port/city interface in African circumstances (Ewais, 1988; Gleave, 1994; Vigarié, 1994), and to the problems posed by the retreat from the waterfront and the consequent need to redevelop older or abandoned port areas (Pirie, 1994). Similarly, there is now an increasing emphasis in port studies on comparative approaches and on the identification of common structures, mechanisms and processes within port systems. While each port city retains its individuality within its own specific geographical, political, economic and technological environments, and as a result develops its own special complexities and problems, any individual cityport nevertheless represents to a greater or lesser extent the overall trends that characterise all such locations and which reflect global rather than local factors.

All cities are continually in transition. A port city reflects in its changing character a wide range of transport-related factors and functions on many scales, from local to global, as well as an equally varied set of urban influences constraining, stimulating or diversifying growth. A recent study has discussed some elements and characteristics of the European port–city system in this context (Hoyle and Pinder, 1992). While many studies of African urban problems make relatively little reference to the port function (East Africa Royal Commission, 1953–55; Hance, 1970; El-Shakhs and Obudho, 1974; Obudho and El-Shakhs, 1979; Obudho, 1981; O'Connor, 1983), others have drawn attention to specific characteristics of port-city growth (Bouthier, 1970; Iliffe, 1970; Hoyle, 1979; Dickinson, 1984; Bahi-Zahiri, 1992). A recent issue of the journal *African Urban Quarterly* has drawn particular attention to port cities (Hoyle, 1994), but overall the literature reveals a broad divergence between port-related and urban-related studies of African port cities. Although the fundamental importance of the port function is recognised, there is implicit recognition also of two quasi-separate circuits of activity, focused upon the port and upon the city, comparable in a sense with Santos' concept of 'the shared space' in the urban economy of developing countries (Santos, 1979).

These aspects of the port/city interface in modern Africa are largely replicated by the cityport/region interface (with which this chapter is primarily concerned). Development planning in modern Africa has tended, on the whole, to consider at the national level an essential series of issues derived from an uneven development surface, an inappropriately structured economy, and a malfunctioning infrastructure. At the regional level, an emphasis on the initiation of growth centres and on economic diversification has been widespread; and in urban terms, an overriding preoccupation has been with the rapid and apparently excessive growth of primate cities alongside the limited and sometimes almost negligible development of smaller urban places. This implies a somewhat compartmentalised view of the development process, defined by region or by economic sector or by location, which seems not to involve a spatial perspective concerned with inter-relationships between city and region or with urban–rural symbiosis as a fundamental element of the day-to-day functioning of cities and regions.

More widely, as Simon (1992) has emphasised, there is now a growing appreciation of African cities as dynamic elements within spatial systems, involved in multidirectional inter-relationships on many scales and in many dimensions, reflecting and responding to, as well as initiating and affecting, the evolving structures and patterns of wider regions. 'It is very difficult to make sense of contemporary Lagos, Abidjan, Dakar or Maputo ... without an understanding of the changing political economy of their respective countries over time and an appreciation of their ... current position within the world system' (Simon, 1992, 4). Although Simon's emphasis here is on national and global spatial scales, the point he seeks to emphasise is equally relevant at the regional scale, as the example of Mombasa illustrates.

MOMBASA AND THE KENYA COASTAL ZONE

Located on the Indian Ocean coast of Kenya (Figures 10.1 and 10.2) – some 400 km long between the borders with Somalia and Tanzania – Mombasa is the most important port city between Port Said (Egypt) and Durban (South Africa) and the leading outlet for an international tributary region some 1.2 million km^2 in extent, inhabited by over 50 million people. The population of Mombasa is very mixed: indigenous Swahili (some of whom claim Persian descent), Arabs, Indians, Africans from coastal and interior Kenya, plus a few Europeans, coexist with their different customs, cultures and creeds. Most are immigrants, attracted to Mombasa by employment opportunities ultimately derived from the port function which has been of continuous and generally increasing importance since the 11th century. Along the eastern coastline of Africa, only Mogadishu (Somalia) can claim the same historical continuity.

Relationships between Mombasa and the Kenyan national space economy are in one sense straightforward, because Mombasa is Kenya's only deep-water port – handling in 1994 some 8.3 million tonnes of cargo – and is therefore a critical element in linking the national and international surface transport systems. In another sense, however, with a fast-growing population approaching 1 million (Sabini, 1994), Mombasa is overwhelmingly the primary urban and industrial node in the coastal zone of Kenya, and its dominance of this zone in socio-economic and political terms yields a wide range of complex issues ranging from water supply and employment to security and industrial decentralisation. The political position of Mombasa in modern Kenya, centred upon the inland capital city of Nairobi, is also somewhat sensitive.

When looking at Mombasa and the coastal zone of Kenya, it is important to distinguish between the various administrative areas involved. Kenya is divided into seven *Regions*, one of which is the *Coastal Region* (83 603 km^2). Within the Coastal Region there are six *Districts* (Mombasa, Kwala, Kilifi, Lamu, Taita and Tana), some of which extend a considerable distance inland; and within *Mombasa District* the *Municipality* of Mombasa is the central urbanised area on the island and neighbouring mainland (Figure 10.2). Two additional geographical concepts transcend these administrative areas. One involves the idea of a *metropolitan* Mombasa, including the Municipality and much of Mombasa District; a second concerns the identification of a *coastal zone* in environmental and ecological terms.

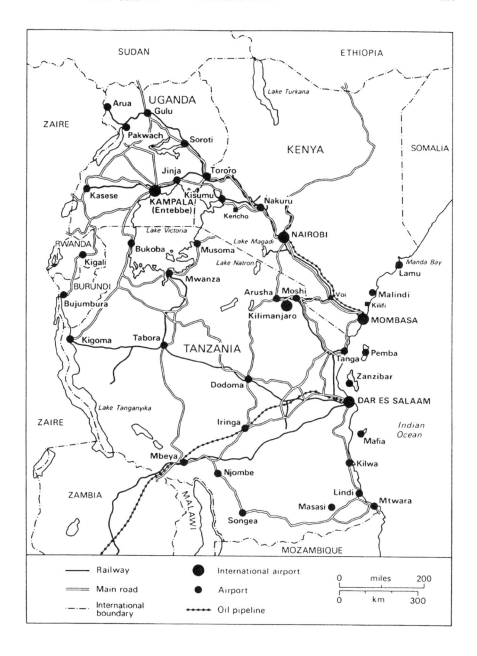

Figure 10.1. The pattern of transport services in East Africa

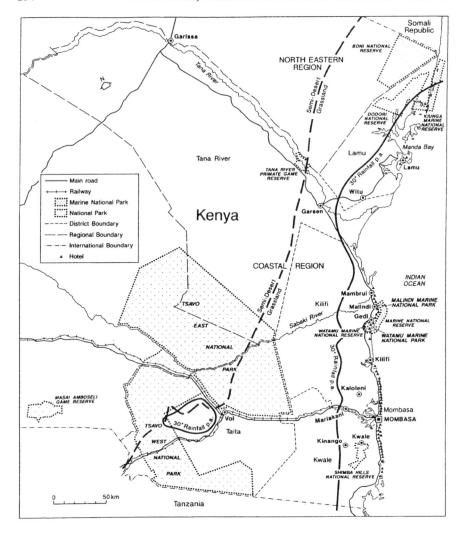

Figure 10.2. The Kenya coastal zone, defined in rainfall/vegetation terms and characterised by urban development and transport axes. Conservation and tourism are important inter-related elements in the zone.

Environmental considerations

The extent to which a cityport can fulfil its developmental role in relation to its local and more distant hinterlands is conditioned by a number of factors. One of these is the physical environment which in the case of Mombasa and the coastal zone is generally advantageous. Defined on one side by the Indian Ocean and on the other by the rapid vegetational transition from the well-watered coastal forest/savanna mosaic to the dry semi-desert commonly known as the *nyika*, the

Kenya coastal zone is a low-lying strip of territory between 10 and 30 miles (16–48 km) wide, with good annual rainfall (over 1000 mm in many places) and some productive soils.

Today, from both the landward and seaward sides, the coastal zone is attractive to settlement, trade and economic development, and as a result is relatively densely populated. The zone is identified in human terms by a mixed settlement pattern which outside Mombasa comprises a variety of small towns (Kilifi, Malindi, Lamu) and smaller communities linked by a north–south road, and an economy based on varied subsistence and commercial agriculture and on tourism. In the past, the seasonal reversal of winds over the western Indian Ocean played a major role in the early development of navigation, trade and settlement along the Kenya coast. From November to March, the north-east monsoon brought sailing vessels from Arabia and India, some of which reached the coast of Tanzania, returning with the build-up of the south-west monsoon in April. From a maritime perspective, the coastal zone presented a relatively productive and not inhospitable environment; but was for a long time largely ignored by traders from the landward side. In fact, early coastal trading settlements were largely cut off from the modern hinterland of Mombasa by the *nyika*, across which there were few reliable routeways before the railway era.

A significant factor from the standpoint of modern port development on the Kenya coast is the existence of a series of drowned river valleys or *rias* resulting from Pleistocene changes in the relative sea level. Some, such as the example of Kilifi, are too shallow or otherwise unsuitable for port development, but at Mombasa *rias* are a critical factor in the water site of the port. Mombasa Island, on which the town centre and many of the port facilities are located, lies between Mombasa Old Harbour and Kilindini Harbour (Figure 10.3). Both of these harbours have been developed for commercial purposes, but in quite different ways. Mombasa Harbour, being rather narrow and relatively shallow, provides shelter for smaller craft, while Kilindini Harbour (the 'place of deep water') provides modern deep-water facilities. The fact that Mombasa has been a seaport of significance in both medieval and modern times is largely due to the geographical juxtaposition on this site of two harbours contrasted in area, depth and capacity. These conditions have enabled the port to adapt itself successfully to functional and navigational changes over time. Moreover, site conditions do not present any insuperable obstacles to physical expansion in the foreseeable future.

Historical perspectives

Today the coastal zone constitutes the maritime façade of a rapidly developing country of considerable economic potential, and forms a vital, outward-looking link with the rest of the world. This orientation is largely a product of the last two hundred years, for during all the long preceding centuries the coastlands were for trading purposes little more than part of the western shore of the Indian Ocean, dependent upon the seasonal reversal of winds. Within this coastal

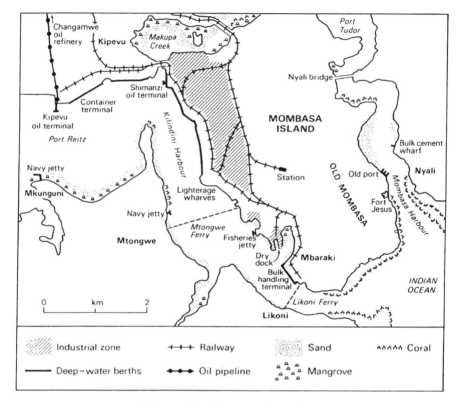

Figure 10.3.   The port of Mombasa

environment successive generations built up widely differing hierarchies of seaports, which comprised important if rather peripheral elements within the widespread network of trading towns and ports stretching in medieval times through south-western Asia to China.

The earliest surviving description of the external trade relations of East Africa is to be found in the *Periplus of the Erythraean Sea* (Huntingford, 1976), a guide to the commerce of the Red Sea and the Indian Ocean written at some time in the first or second century AD by a Greek trader living in Alexandria. The book describes a voyage along the East African coast and mentions, *inter alia*, the trading port of Rhapta which was subsequently described as a 'metropolis' in Ptolemy's *Geographia* (Stevenson, 1932). The site of Rhapta has never been satisfactorily identified. It used to be associated with the Rufiji delta in Tanzania, but recent work by Dr Mark Horton of the British Institute in Eastern Africa suggests that Manda Bay on the north Kenya coast may be a more likely site.

Early medieval trading ports, such as Manda and Mombasa, to which Lamu and Malindi were soon added, probably dominated in turn from their defensive sites varying areas of coast and hinterland, their fluctuating comparative importance reflecting their varying fortunes in trade and warfare. The experience of Kilwa, the principal medieval port on the coast of Tanzania, was essentially similar. Mombasa Island is known to have served as a maritime trading post in the 11th century, and it began to take shape as a town with the Shirazi migrations (from southern Arabia and southern Persia) in the 13th century. Surprisingly, perhaps, there has been very little archaeological investigation on Mombasa Island. We know that there was a settlement from the 11th century onwards, but we cannot rule out the possibility that a town or a port existed at an earlier date. As Richard Burton put it, 'it is hard to believe that the Phoenician, Egyptian and Greek merchants would have neglected the finest harbour and the best site for trade upon the whole Azanian coast' (Burton, 1872, Vol. 2, 37).

The later middle ages were marked by a much fuller development of Islamic civilisation, with rapid urban expansion and trade development especially in the 14th century, possibly associated with improved environmental and political conditions (Stiles, 1992). Among settlements from this period along the Kenya coast that have not survived but have been excavated, the best preserved is Gedi, located 13 km south of Malindi. Gedi was not a seaport, but illustrated the character of the Arab colonial towns at the height of their prosperity. Founded in the late 13th or early 14th century, Gedi reached its apogee in the mid-15th century and was finally abandoned in the early 17th century. Little is known of its history but evidence suggests that it had a large and relatively prosperous population, perhaps exceeding 15 000 (Kirkman, 1954, 1956, 1964).

The full development of Arab settlements and their trade and culture on the Kenya coast in the 15th century, dominated by Mombasa, immediately preceded a period of decline. At the end of that century, and from the south, 'the restless energy of western Europe intruded upon the East African coast like an unseasonable monsoon for which the inhabitants were totally unprepared' (Ingham, 1962, 6). The Portuguese programme of African coastal exploration culminated in the celebrated voyage of Vasco da Gama to India in 1497–99. He was impressed by Mombasa, but received a warmer welcome at Malindi, where he erected a stone cross and found a pilot to guide him across to Calicut. A few years later, in 1517, the Portuguese navigator Duarte Barbosa noted a degree of port–city interdependence at Mombasa:

> Further on ... there is an isle hard by the mainland, on which is a town called Mombaça. It is a very fair place, with lofty stone and mortar houses, well aligned in streets (after the fashion of Quiloa) ... This is a place of great traffic, and has a good harbour, in which are always moored craft of many kinds and also great ships ...   (Dames, 1918, 1921)

Soon the Shirazi town, on the eastern side of the island facing the Old Mombasa Harbour, was paralleled and eventually superseded by the Portuguese town of the 16th and 17th centuries. Although the present urban structure of the old town of Mombasa was largely shaped during Portuguese times, the overall effect of Portuguese intervention on the Kenya coast was negative, and Mombasa maintained an attitude of open revolt against their authority from the time of da Gama's first arrival in 1497 until their final withdrawal to the south early in the 18th century. The impressive, formidable Fort Jesus stands now as the only substantial physical monument to their rule (Kirkman, 1964). Rezende's map of Mombasa, dated 1636, now in the British Museum in London, shows the relationships between the fort, the Arab town, the island and the two harbours, although the size of the fort is grossly exaggerated (Figure 10.4).

The rising tide of Arab-controlled slave trading in 19th-century East Africa severely disrupted the economic and social fabric of the area (Nicholls, 1971). Selected by the Omani Arabs as a regional emporium, the offshore island of Zanzibar (today part of Tanzania) was the chief 19th-century centre of innovation. The re-entry of Europeans on the East African scene in the later 19th century coincided with important technological changes: the opening of the Suez Canal (1869), the change from sail to steam as a means of propulsion of vessels, the rapidly increasing size of ships, and the growing importance of railways. A combined result of these innovations was that in early colonial East Africa arterial railways were built from selected port sites (such as Mombasa) chosen for their ability to accommodate larger steamers in a context of increasing trade with Europe via Suez.

In Kenya, and indeed in eastern Africa generally, Mombasa was the principal beneficiary in this process whereby a traditionally fluid port pattern became crystallised as more capacious, sheltered, deep-water harbours replaced the

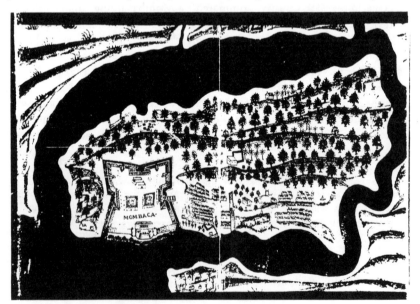

Figure 10.4. A plan of Mombasa in 1636, from Boccaro Rezende's *Livro da Estado da India.* (original held in British Museum, London)

minor inlets and open shorelines previously used by smaller ships. The intention to build railways, themselves powerful agents of innovation and economic transformation in the hinterlands, was the immediate cause of concentration of interest and activity on appropriate modern seaports, a process that took place in many colonial territories. Mombasa was particularly fortunate at this point in time, for its general geographical location and its specific site conditions enabled the Kenyan port city to establish and maintain a central place in the modern economic life of the coastal zone.

Political geography

The coastal zone provided a problem of political geography that involved Mombasa and other East African seaports during the European colonial period (1880s–1960s). The East African coastal strip was defined, politically, as the zone lying approximately within 10 miles of the coast. For centuries the coastlands and offshore islands of East Africa were largely controlled by immigrant Arab traders, latterly through the Sultan of Zanzibar, and there existed for a long time a considerable degree of geographical unity, physical and human, reflected in historical circumstances and population conditions. Under British colonial rule, however, the Zanzibar islands became a protectorate, whereas Kenya became a colony and Tanganyika (after 1919) a mandated territory.

With the coming of the Europeans to East Africa in larger numbers during the 19th century, the southern section of the strip (now in Tanzania) was ceded by the Sultan of Zanzibar to Germany in 1890, and thus became part of German East Africa and in 1919 of Tanganyika. The Kenya section, however, administered by Britain, remained technically under the authority of the Sultan until 1963 (Melamid, 1963). The situation of Mombasa within the Kenya section of the strip added point to the need to find a solution to the problem prior to Kenya's independence in 1963. A commission of enquiry reported that a majority of the inhabitants of the strip favoured union with Kenya (Robertson, 1961). The *de jure* Arab rule of the coastal strip was thus brought to a peaceful end; but this was quickly followed, early in 1964, a few weeks after Zanzibar's independence, by a political revolution against the new Arab government which also ended the *de facto* Arab rule in the former protectorate.

The rise of the modern port

The point of departure for the modern port of Mombasa was the purchase in 1895 of land near Kilindini Harbour as a base from which to direct the building of the railway through interior Kenya to Lake Victoria. Using Mombasa as an initial base, Britain had assumed political control of Kenya in 1895, and (as elsewhere) a standard procedure was to consolidate that control with an outline transport infrastructure, beginning with a railway to the interior from a selected port site (Hill, 1949). Mombasa was the obvious choice, as an established town, with a deep-water harbour of recognised potential. Earlier, Captain Owen had written that 'Perhaps there is not a more perfect harbour in the world than Mombasa' in his account of the 1820s hydrographic survey of the East African coast (Owen, 1833, 412). Seventy years later, information was translated into action. 'The port possesses great facilities for development as well as sites for warehouses and wharves of almost indefinite extension' (Molesworth, 1899) '. . . The most urgent necessity is a deep-water berth . . . (and) a comprehensive plan showing what will be the ultimate aim and object to be attained when traffic largely develops' (Gracey, 1901).

From these beginnings, the port of Mombasa has grown throughout the 20th century, beginning with lighterage wharves at Mbaraki and proceeding upstream from 1926 to 1958 with deep-water berths along the north-western shore of the island. From the 1960s to the present day, additional deep-water berths and various forms of specialised quayage including a container terminal and oil-reception facilities have been added on the mainland at Kipevu. There are plans to extend these facilities, in the context of a maritime industrial development area, along the southern side of Port Reitz. The urban area, for long confined to the island, has gradually spread to the adjacent mainland on both the northern and southern sides and also north-westwards along the road/rail axis towards Nairobi.

The second port debate

The port function remains central to the urban and regional economy of Mombasa and the coastal zone, but there is some anxiety that too high a proportion of economic activity is concentrated within Mombasa Municipality and District. In theory, a national or regional port hierarchy is a dynamic phenomenon which may gain or lose constituent elements in the course of time. This has happened on the Kenya coast in the past, but not during the 20th century. Elsewhere, new ports of various kinds have been established in several other African countries in recent years, often in the context of a major regional development programme, as at Tema (Ghana) (Hilling, 1966) and Richards Bay (South Africa) (Weise, 1981 and 1984). In Kenya the possibility of establishing a second deep-water port as a basis for a new urban and industrial growth pole within the coastal zone has been under discussion since the 1970s (Republic of Kenya, 1977). The virtually complete dependence of Kenya upon the port of Mombasa is a sensitive issue in political and strategic terms, and the creation of alternative coastal growth centres would accord with the Kenya government's broad strategy of development diffusion.

The site selected for detailed consideration is at Manda Bay, on the north Kenya coast near the historic island port of Lamu (Figure 10.5). The site offers a

---

Figure 10.5. (*opposite*) The proposed port of Manda Bay, Kenya

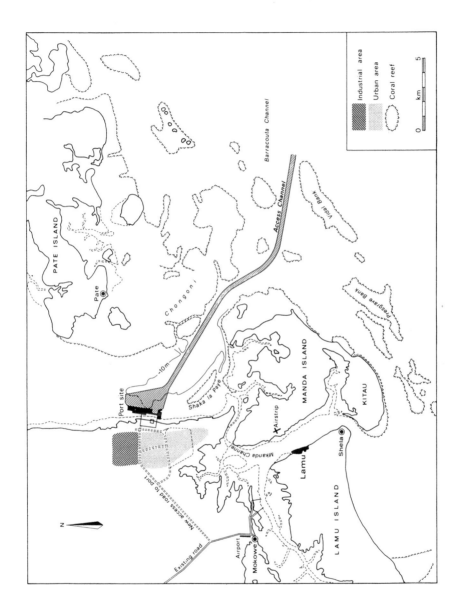

splendid, sheltered deep-water harbour and extensive areas of level land for development where a new industrial cityport could be created at relatively low cost. Provided that adequate hinterland communications were established – clearly a vitally important condition – Manda Bay could become a basis for a new growth pole of considerable long-term potential. Although the arguments in favour of such an innovation seem attractive, those against have hitherto proved more powerful. In the absence of any large-scale industrial traffic generator, it seems likely that the industrial prospects of a new port would be limited. The possible negative effects of a major port–industrial zone, close to the world-renowned tourist centres along the coastal zone, notably Lamu, must also provide a disincentive to development.

In this respect the proposed port contrasts with most other new cityport growth poles in modern Africa, where a specific industrial *raison d'être* has provided an essential stimulus for port development. The processing of iron ore at Saldanha Bay (South Africa), Nouadhibou (Mauritania) and Buchanan (Liberia), the treatment of timber at San Pédro (Côte d'Ivoire), the aluminium industry at Tema (Ghana) and the phosphate developments at Lomé (Togo) all provide examples of motivating industries without parallel in the Kenya case. The arguments in favour of Manda Bay rest primarily on the socio-economic and strategic desirability of developing a neglected corner of the national territory – an area of considerable long-term potential but limited resource endowment – and also to some extent on the perceived need to provide a solution to actual and potential problems associated with port and urban congestion at Mombasa as Kenya's principal maritime industrial development area continues to expand.

Urban and regional planning

The question of a second deep-water port was debated at an international seminar held in Mombasa in 1992 to review issues involved in urban and regional planning in the coastal zone. The conference took the view that the question should be a matter of continuing, open debate. On balance, the construction of such a port (and associated urban and industrial areas) is desirable in terms of broad, long-term strategies, but questionable in terms of more immediate, specific problems. 'We therefore recommend that the matter be kept under careful and continuous review, with special reference to the precise locations under consideration, and to the eventual integration of any selected site within the Kenyan space economy as a whole' (Obudho, 1992, 26).

A number of further conclusions and recommendations from the 1992 Mombasa seminar deserve emphasis, some of which concern specifically the urban area, but most of which relate to the coastal zone as a whole and its regional and national dimensions and relationships. There is clearly a perceived need to clarify relationships between Mombasa and its region, distinguishing between administrative units and geographical areas. There is some confusion between the municipal, district and regional areas, and with the concept of a metropolitan urban–industrial zone. There is a need for, and an increasing awareness of, a closer scrutiny of relationships between port and city, and between cityport and region.

Among the most critical issues of concern to those who live and work in the coastal zone, and those who have the responsibility for designing planning strategies, the specific problem of assured supplies of water of good quality is paramount. This question links together the context of increased pressures in a rapidly expanding city where housing policy is inevitably a major issue; problems of pollution derived from port, industrial and urban activities; and a generally increasing awareness of environmental questions and responsibilities. Beyond this, one of the most complex questions in both urban and rural parts of the coastal zone concerns land occupancy and land tenure, invariably a difficult problem in a multicultural society with deep roots and a strong attachment to land resources. Infrastructural problems of increasing concern include the provision of modern, efficient transport systems; as a whole, the coastal zone is over-reliant upon road transport, notably on private cars and on bus services that are sometimes less than efficient. Urban and local rural roads have been improved greatly in recent years, but still constrain mobility. The major axis parallel to the coast, however, still relies on ferries to link Mombasa with the southernmost part of the coastal zone (a high-level bridge has been discussed but rejected on grounds of cost); and the major national transport axis is, of course, the Mombasa–Nairobi route used by rail, road and pipelines (Republic of Kenya, 1989a and b).

The development of the Kenya coastal zone as a major national and international tourist destination area has brought many benefits in terms of infrastructural provision – roads, bridges (as at Kilifi), local air services – and an increased environmental awareness expressed, for example, in the establishment of a marine national park at Watamu (near Malindi) and in the close attention now being given to the conservation of the urban fabric of the traditional core areas of Mombasa, Malindi and Lamu (*A conservation plan ...*, 1990). Tourism is regarded throughout the coastal zone, as indeed throughout Kenya, as a very valuable if rather sensitive component of the economy. It would be difficult to overemphasise the importance of tourism to the Kenyan economy, for in recent years this industry has replaced coffee exports as the country's principal earner of foreign exchange. There is, however, a continuing need for increased harmonisation between the demands of tourism and the needs of other elements in the local economy.

Mombasa is conscious, as a major cityport, of its position within the Kenya coastal zone and within the national Kenyan economy. Like many modern cityports of colonial origin and development, Mombasa was once a political capital. From 1895, when British control over what is now Kenya was firmly established, to 1907, two years after the railway from Mombasa to Lake Victoria had been completed, Mombasa served as the colonial political capital. It became clear, however, that a more centrally located capital would serve far better the interests of the colony as a whole, and especially the interests of the European settlers in the highlands of the interior. Mombasa therefore quickly lost its national capital status, and has subsequently been regarded by Nairobi as a useful functional appendage. Today Mombasa believes, with some justification, that it is somewhat undervalued within the Kenyan state. As a cityport of national and international significance, Mombasa has developed an international airport, but has not yet been accorded city status and has not yet acquired a university, both of which are long overdue in view of the substantial and diverse demographic and economic base of the municipality and district.

The Kenya coastal zone illustrates on a local scale the widespread problem of imbalanced urban hierarchies in African and other less developed countries. Within Kenya, Mombasa is second in size only to Nairobi, the capital. Within eastern Africa, Mombasa is an urban centre of the first importance. Within the coastal zone, all other urban places (Malindi, Lamu) are virtually insignificant by comparison. In such circumstances the conventional urban/rural divide may be rejected in favour of a more realistic dichotomy between the largest urban places, on the one hand, and the smaller urban centres and rural areas on the other. The notion of controlling city size can be rejected as impracticable, given that modern world society is increasingly urbanised; but the desirability of fostering the growth of medium-sized and smaller towns and growth centres should be emphasised so as to encourage the emergence of a more varied and balanced settlement hierarchy within the space economy of regions and nations.

CONCLUSIONS

In conclusion, several aspects of the inter-relationships between cityports, coastal zones and African development may be underlined. The impressive expansion of individual port cities which has occurred during the 20th century has played a vital role in promoting mutually beneficial interlinkages between African countries and the global economy. The port system has made available to world consumers a variety of resources essential to economic growth. The bulk transport of minerals, the export of agricultural produce and the dramatic rise in energy imports, notably crude oil, have become widespread components of African port throughput structures.

These trends have not affected the African cityport system evenly. Some countries, cities and port authorities have made impressive investments in infrastructure, making their ports attractive to new generations of ships, and cityport growth poles of this kind have become especially important for African development. The quality of transport links between ports, regions and hinterlands have also, of course, been highly significant. Those cityports enjoying the best rail, road and pipeline access have been able to attract an increasing proportion of maritime traffic, and have also been able to promote the introduction of interior container depots (ICDs) as a means of improving hinterland transport efficiency. The implementation of efficient road/rail communications between ports and hinterlands is a necessary planning strategy for developing a more decentralised port system throughout the continent. The African cityport system is a dynamic phenomenon, constantly changing in response to a variety of factors on the landward and seaward sides. Through the centuries the essential pacemakers have been the maritime factors. New technologies of ship design and cargo handling have led to successive eras of cityport evolution, and modern cityports have become an integral and important part of the continent's economy, at many different scales. Africa's seaports and cityports are continually involved, however, in a process of inter-port competition. Modern seaports, like those in the past, survive by attracting traffic; those which fail to do so decline or stagnate. Decision-making by ship operators and hinterland transport companies, as well as by commodity producers and consumers, yields a port selection process with advantages for some port cities but disadvantages for others. Political factors, too, affect the ways in which ports are used, in numerous ways: the complex surface transport systems which Africa has inherited from the colonial past continue to shape present-day traffic flows; governments of inland states manipulate commodity flows in their anxiety to preserve access to ocean ports; and instability and warfare, notably in Angola and Somalia, continue to prevent the use of some seaports for normal economic purposes.

Another negative factor of widespread importance is the relatively weak state of most African national economies. Although a few tropical African countries, notably Nigeria and Zaire, are sometimes cited as being among the potential industrial giants of the 21st century, their enormous resources are of little value without internal political stability and orderly economic development planning. Other, much smaller countries, such as Côte d'Ivoire and Kenya, have made substantial economic strides but, as they aggravate their highly uneven development surfaces, their apparent prosperity remains fragile. The emerging new South Africa may yet become the economic salvation of the continent. At present, however, political instability is the prime cause of slow economic progress, for capital investment will not venture where it may be misused, and Marxist approaches have blatantly failed. The economic condition of the continent as a whole is such that, as far as port cities and their coastal zones in the developing world are concerned, no African location has yet emerged to rival the cityport emporia of Latin America or the New Pacific, and no African country (with the possible exception of South Africa) can yet claim to be a newly industrialising country (NIC) in the full sense in which that term is understood in southern and eastern Asia. 'The absence of any true African NICs and hence dynamic cities operating within the system of core cities which controls international circuits of capital, commodities and skilled labour, such as Singapore, Hong Kong or São Paulo, is clearly related to Africa's global economic peripherality ... Despite the growth of international communications ... Africa is for the most part becoming relatively even more peripheral at present' (Simon, 1992, 4).

If this is so, then the outlook for African port cities and their regions in the immediate future is not very bright. In the longer run, things may improve. It is perhaps not without significance that in 1995 the Association Villes et Ports held its 5th International Conference in Dakar (Sénégal), to focus especially on the role of port cities in regional development in the less developed countries. Meanwhile, from a physical planning perspective, perhaps the most critical issue affecting the African cityport today is the overconcentration of port activity in a single national port city, or at best in a limited hierarchy of port cities. Africa has long suffered from a tendency towards the overconcentration of urban growth in a very limited number of large centres; and the tendency towards port concentration is enhanced not only by established African urban trends but also by global maritime transport factors. Although some African countries have successfully introduced new ports for specific purposes or in a context of regional development, it remains very difficult for African governments and their financial advisors and supporters to move against established trends towards concentration. This problem of geographical concentration versus dispersion of port-city growth, and more widely of economic development as a whole, remains a critical issue for development planners in African countries today, and strategies adopted for its solution will profoundly affect the character of Africa's systems of cityports and cityport regions in the future.

REFERENCES

A conservation plan for the Old Town of Mombasa, Kenya (1990), prepared for the National Museums of Kenya and the Municipal Council of Mombasa, with financial and technical assistance from the UNDP and UNESCO.
Bahi-Zahiri, M. (1992), 'Urban transport in Abidjan', Aquapolis 1(6), 28–31.
Bouthier, M. (1970), 'The development of the port of Abidjan and the economic growth of Ivory Coast', in Hoyle, B.S. and Hilling, D. (eds), Seaports and development in tropical Africa (London: Macmillan), 103–26.
Breen, A. and Rigby, D. (1993), Cities reclaim their edge (Washington, DC: Waterfront Center).
Broeze, F., Reeves, P. and McPherson, K. (1986), 'Imperial ports and the modern world economy: the case of the Indian Ocean', Journal of Transport History 7, 1–20.
Burton, Sir R. (1872), Zanzibar: city, island and coast (London: Tinsley Bros).
Christopher, A.J. (1985), 'Continuity and change of African capitals', Geographical Review 75, 44–57.
Dames, M. L. (ed.) (1918, 1921), The book of Duarte Barbosa (London: Hahluyt Society), 2 vols.
Dickinson, B. (1984), 'The development of the Nigerian ports system: crisis management in response to rapid economic change', in Hoyle, B.S. and Hilling, D. (eds), Seaport systems and spatial change: technology, industry and development strategies (Chichester: Wiley), 161–77.
East Africa Royal Commission (1953–55) Report (London: HMSO, CMd 9475).
El-Shakhs, S. and Obudho, R.A. (eds) (1974), Urbanization, national development and regional planning in Africa (New York: Praeger).
Ewais, H.M.H. (1988), 'Land use policies in a cityport, with special reference to Alexandria', University of Liverpool, unpublished PhD thesis.

Forjaz, J. (1992), 'Sea, port and city: three inseparable terms in the urban history of Maputo', *Aquapolis* 1(3), 14–19.

Gillman, C. (1945), 'Dar es Salaam, 1860–1940: a story of growth and change', *Tanganyika Notes and Records* 20, 1–23.

Gleave, M.B. (1994), 'Freetown, Sierra Leone: port activities, urban growth and urban area differentiation', *African Urban Quarterly* 9(1,2) (in press).

Gracey, T. (1901), 'Correspondence respecting the Uganda railway', *Africa* 6 (cited in Hill, M.F., 1949, 204–10).

Griffiths, I. (1990), 'The quest for independent access to the sea in southern Africa', *Geographical Journal* 155(3), 378–91.

Hance, W.A. (1970), *Population, migration and urbanization in Africa* (New York: Columbia University Press).

Hill, M.F. (1949), *Permanent way: the story of the Kenya and Uganda Railway* (Nairobi: East African Railways and Harbours).

Hilling, D. (1966), 'Tema, the geography of a new port', *Geography* 51, 111–25.

Hoyle, B.S. (1967), 'Early port development in East Africa: an illustration of the concept of changing port hierarchies', *Tijdschrift voor Economische en Sociale Geografie* 58, 94–102.

Hoyle, B.S. (1972), 'The port function in the urban development of tropical Africa', *La croissance urbaine en Afrique noire et à Madagascar* (Colloques Internationaux du Centre National de la Recherche Scientifique, 539, Talence 1970), II, 705–18.

Hoyle, B. S. (1979), 'African socialism and urban development: the relocation of the Tanzanian capital', *Tijdschrift voor Economische en Sociale Geografie* 70, 207–16.

Hoyle, B.S. (1981), 'Cityport industrialization and regional development in less-developed countries: the tropical African experience', in Hoyle, B.S. and Pinder, D.A. (eds) *Cityport industrialization and regional development: spatial analysis and planning strategies* (Oxford: Pergamon Press, Urban and Regional Planning Series, Vol. 24), 281–303.

Hoyle, B.S. (1983), *Seaports and development: the experience of Kenya and Tanzania* (New York and London: Gordon and Breach).

Hoyle, B.S. (1985), 'Gateways from the sea: some maritime perspectives on African seaports', in Stone, J.C. (ed.) *Africa and the sea* (Aberdeen: University of Aberdeen African Studies Group), 352–77.

Hoyle, B.S. (1988), *Transport and development in tropical Africa* (London: John Murray).

Hoyle, B.S. (1989), 'Maritime perspectives on ports and port systems: the case of East Africa', in Broeze, F. (ed.) *Brides of the sea: port cities of Asia from the 16th to the 20th centuries* (Sydney: New South Wales University Press; and Honolulu: University of Hawaii Press), 188–206.

Hoyle, B.S. (1994), 'The changing port city in modern Africa', *African Urban Quarterly* 9(1,2) (in press).

Hoyle, B.S. and Hilling, D. (eds) (1970), *Seaports and development in tropical Africa* (London: Macmillan).

Hoyle, B.S. and Hilling, D. (eds) (1984), *Seaport systems and spatial change: technology, industry and development strategies* (Chichester: Wiley).

Hoyle, B.S. and Knowles, R.D. (eds) (1992), *Modern transport geography* (London: Belhaven, for the Transport Geography Study Group, Institute of British Geographers).

Hoyle, B.S. and Pinder, D.A. (eds) (1992), *European port cities in transition* (London: Belhaven Press, in association with the British Association for the Advancement of Science).

Hoyle, B.S., Pinder, D.A. and Husain, M.S. (eds) (1988), *Revitalising the waterfront: international dimensions of dockland redevelopment* (London: Belhaven).

Huntingford, G.W.B. (1976), *The Periplus of the Erythraean Sea* (London: Hakluyt

Society, 2nd series, vol. 151).

Iliffe, J. (1970), 'A history of the dockworkers at Dar es Salaam', *Tanzania Notes and Records* 71, 119–48.

Ingham, J. (1962), *A history of East Africa* (London: Longman).

Kirkman, J.S. (1954), *The Arab city of Gedi: excavations at the great Mosque* (London: Oxford University Press).

Kirkman, J.S. (1956), 'The culture of the Kenya coast in the later middle ages', *South African Archaeological Bulletin* 11, 89–99.

Kirkman, J.S. (1964), *Men and monuments on the East African coast* (London: Lutterworth Press).

Kironde, J. M. L. (1993), 'Will Dodoma ever be the new capital of Tanzania?' *Geoforum* 24(4), 435–53.

Mabogunje, A.L. (1989), *The development process: a spatial perspective*, 2nd edn (London: Unwin Hyman).

Melamid, A. (1963), 'The Kenya coastal strip', *Geographical Review* 53, 457–9.

Molesworth, Sir G. (1899), 'Report on the Uganda Railway', *Africa* 5 (cited in Hill, M.F., 1949, 179–85).

Moorsom, R. (1984), *Walvis Bay: Namibia's port* (London: International Defence and Aid Fund for Southern Africa, in cooperation with the United Nationals Council for Namibia).

Nicholls, C.S. (1971), *The Swahili coast: politics, diplomacy and trade on the East African littoral, 1798–1856* (London: Allen & Unwin).

Obudho, R.A. (ed.) (1981), *Urbanisation and development planning in Kenya* (Nairobi: East African Literature Bureau).

Obudho, R.A. (1992), *Urban and regional planning of Mombasa and the Coastal Region, Kenya: report of an international workshop* (Nairobi: Urban Centre for Research, and Institut Français de Recherche en Afrique).

Obudho, R.A. and El-Shakhs, S. (eds) (1979), *Development of urban systems in Africa* (New York: Praeger).

O'Connor, A.M. (1983), *The African city* (London: Hutchinson).

Owen, Capt. W.F.W. (1833), *Narratives of voyages to explore the shores of Africa, Arabia and Madagascar* (London: Bentley, 2 vols).

Pirie, G.H. (1994), 'Revitalising South African docklands', *African Urban Quarterly* 9(1,2) (in press).

Republic of Kenya (1977), *Manda Bay port: feasibility study of Kenya's second port* (Nairobi: Renardet-Sauti for Ministry of Power and Communications).

Republic of Kenya (1989a), *Development plan, 1989–93* (Nairobi: Government Printer).

Republic of Kenya (1989b), *Mombasa District Development Plan, 1989–93* (Nairobi: Government Printer).

Robertson, J.W. (1961), *The Kenya coastal strip: report of the Commissioner* (London: HMSO, Cmnd 1585).

Sabini, M. (1994), 'Port and urban form: the case of Mombasa Kenya', *African Urban Quarterly* 9(1,2) (in press).

Santos, M. (1979), *The shared space: the two circuits of the urban economy in underdeveloped countries* (London: Methuen).

Schulze, W. (1970), 'The ports of Liberia: economic significance and development problems', in Hoyle, B.S. and Hilling, D. (eds), *Seaports and development in tropical Africa* (London: Macmillan), 75–101.

Simon, D. (1992), *Cities, capital and development: African cities in the world economy* (London: Belhaven).

Stevenson, E.L. (1932), *The geography of Claudius Ptolemy* (New York).

Stiles, D. (1992), 'The ports of east Africa, the Comoros and Madagascar: their place in

Indian Ocean Trade from 1–1500 AD', *Kenya Past and Present* **24**, 27–36.

Stone, J.C. (ed.) (1985), *Africa and the sea* (Aberdeen: University of Aberdeen African Studies Group).

Vigarié, A. (1994), 'Abidjan, Côte d'Ivoire: le port et son rôle dans le développement urbain', *African Urban Quarterly* **9**(1,2) (in press).

Weise, B. (1981), *Seaports and port cities of southern Africa* (Wiesbaden: Franz Steiner Verlag GMBH, Kölner Geographische Arbeiten, 11).

Weise, B. (1984), 'The role of seaports in the industrial decentralisation process: the case of South Africa', in Hoyle, B.S. and Hilling, D. (eds), *Seaport systems and spatial change: technology, industry and development strategies* (Chichester: Wiley), 415–34.

11 Diversifying the Cityport and Coastal Zone Economy: The Role of Tourism

ANDREW CHURCH
Department of Geography, Birkbeck College, London, UK

In the 1980s and 1990s a variety of agencies responsible for economic development in cityport regions have sought to diversify the local economy through policies seeking to promote the tourist industry and attract visitor expenditure from outside the region. Tourist strategies in such regions have often been very wide-ranging and have included initiatives for the heritage, arts, culture, entertainment, leisure and hospitality industries. In some of Britain's major waterfront regeneration locations key flagship redevelopment projects are presented as tourism initiatives. For example, despite their very different functions the Tate Art Gallery in the Albert Dock in Liverpool, the London Arena (a sport and music venue in London Docklands), and the naval heritage centre in Portsmouth are all promoted as tourism-orientated developments.

Equally, the similarities between certain waterfront tourism regeneration measures are well known. Marinas, specialist shopping arcades and historic ships, original or otherwise, are common elements of port redevelopment in cityport regions. It is tempting to suggest that many former cityports and cityport regions are trying to pull the same development trick as, for example, the successful models of Baltimore or Boston, USA. A heritage-based 'people draw' attraction accompanied by high-margin specialist shops and appropriate theme restaurants is developed initially to bring in visitors and promote investor confidence. Public money further stimulates private sector interest through site clearance, environmental improvements, new infrastructure and image reconstruction. The momentum created leads to integrated office and private housing development on neighbouring sites.

Such a summary of the waterfront regeneration process is of necessity an oversimplification as there is much local variation. However, the similar role that emerges for tourist and heritage attractions has often been noted. Boyer (1992), in a textual analysis of historic preservation and the symbolism of the built

Cityports, Coastal Zones and Regional Change. Edited by Brian Hoyle.
© 1996 John Wiley & Sons Ltd.

environment at South Street Seaport, New York, concludes that 'City after city discovers its abandoned industrial waterfront or outmoded city centre contains enormous tourist potential and refurbishes it as a leisure-time spectacle and sightseeing promenade. All of these sites become culinary and ornamental landscapes through which the tourists – the new public of the late twentieth century – graze, celebrating the consumption of place and architecture, and the taste of history and food' (Boyer, 1992, 189).

The identification of similarities in the role of tourism at high-profile waterfront sites is clearly useful both for understanding urban change and for policy analysis. It is important to remember, however, that similarity of outcome will not always reflect similarity of intention by the public agencies involved. Furthermore, when considering a more extensive cityport region, comparisons may be harder to identify. A cityport region will contain a complex mosaic of current and potential tourism developments which may or may not relate to each other. In addition, a range of agencies will be operating, all subject to differing political, economic and organisational influences. In such a situation the position of tourism on the policy agenda, in the local economy and in local socio-spatial structures may be quite varied. This chapter considers the nature of tourism policy, its consequences and variations, in three cityport regions in the UK: Dover, Teesside/Hartlepool and London Docklands (Figure 11.1). The contrasts that emerge suggest that despite the seeming ubiquity of certain types of tourism consumption, local agencies do have a certain degree of choice over the type of approach adopted towards tourism in a cityport economy.

UNDERSTANDING TOURISM AND TOURISM POLICY IN THE CITYPORT ECONOMY – CONCEPTUAL ISSUES

Textual readings of the tourist waterfront landscape, such as that by Boyer (1992) mentioned above, indicate that cityport economic development initiatives based on tourism require a complex political, cultural and economic interpretation. Not only do tourism initiatives represent an act of private and/or public capital investment seeking a return from visitor expenditure, but they also serve a symbolic function that may legitimise many other socio-economic elements of post-industrial society. Indeed, the economic and social significance of tourism leads Urry (1993) to suggest that it is 'as important a feature of modern western societies as is the social production of manufactured goods' (p. 91).

Recent studies have sought to examine the production and consumption of tourism in advanced Western countries from a range of theoretical perspectives. A geographical analysis of tourism by Shaw and Williams (1994) argues that the current production of tourism is typified by five main features: the commodification of leisure; the privatisation of services; the interdependence of tourism and leisure; state intervention; and a production process based on the formal economy, informal production and households. Such a production process involves the social construction of the objects of tourism. Urry (1990)

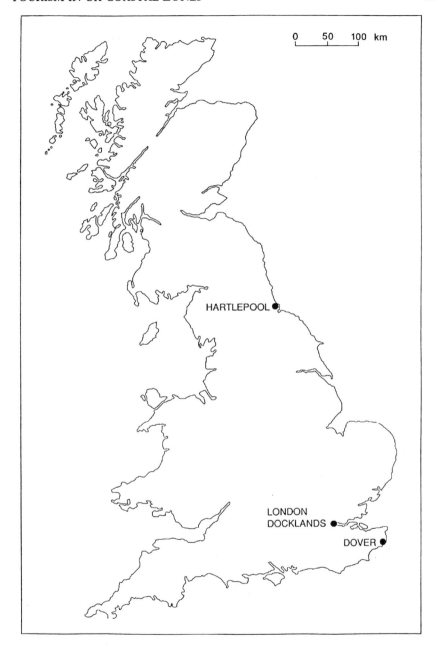

Figure 11.1. The location of three case study areas

argues that tourism consumption can be summarised as a 'gaze' as consumers increasingly demand 'out of the ordinary' experiences. The 'gaze' contributes to the process of developing personal social identity in an increasingly disorienting, fragmented and internationalised society and economy. The social theory perspective suggests, therefore, that tourism has an increasingly important social and cultural role as spaces become remade to play a part in the tourist 'gaze'. Such a process may involve altering the existing space/power relations so that their tourists feel a sense of 'ownership'. This may be disadvantageous to, for example, local residents and Bondi (1992) also notes the consequences of new forms of urban spaces for gender relations. In addition, the social construction of tourism products may involve the reinterpretation of the role spaces play in memory (Urry, 1993). The need for the heritage industry to create brief, easily accessible and acceptable versions of history may well conflict with the memories of those involved in the past activities at a particular location (Hewison, 1987).

The process by which spaces are remade for tourism production and consumption requires an understanding not only of economic and political processes, but also an appreciation of the process by which spatial changes are encoded in symbolic images as well as the built environment. Sorkin (1992) argues that contemporary city-centre renewal, waterfront redevelopment and gentrification provide 'urban renewal with a sinister twist, an architecture of deception which, in its happy-face familiarity, constantly distances itself from the most fundamental realities. The architecture in this city is almost purely semiotic, playing the game of grafted signification, theme-park building' (p. xiv). Some analyses stress the negative side of this process stemming from the individualistic and private nature of new urban cultural and tourist experiences (Booth and Boyle, 1994). Others, such as Getz (1989), argue that certain tourist events, such as large-scale heritage celebrations, can promote social benefits in terms of a sense of belonging, authentic historical understanding and the involvement in spectacle, ritual and games. Whatever view is taken of this process it is apparent that an understanding of the development of tourism policy in cityport regions requires an analysis that recognises the diverse economic, social, and cultural influences on the production and consumption of tourism products.

The involvement of the state in tourism policy is a comparatively recent phenomenon given the relatively long history of the industry in the United Kingdom (Pearce, 1992) and until the 1980s national tourism policy has been portrayed as limited and lacking direction (Cooper, 1987; Shaw et al., 1988). General discussions of the state's role in tourism policy have considered the characteristics of policy evolution both in the United Kingdom and elsewhere. In a comparative analysis Pearce (1992) identifies marketing, economic development, planning, visitor servicing, research, coordination and lobbying as the dominant roles of tourism organisations. Pearce adopts an organisational perspective and argues that the key reasons for the formation of tourism organisations include the interdependence of sectors, the small size of firms, market fragmentation and the spatial separation of origins and destinations. The discussion of state tourism policy by Shaw and Williams (1994) provides a more political interpretation and outlines six reasons for state interventionism: economic goals; political legitimation; equity and social needs; externalities and social investment; regulation and negative controls; and regional development. The changing influence of such factors over time has been examined by studies of United Kingdom tourism policy at the national level and these have identified the importance of national economic goals, such as the desire to boost foreign exchange earnings, as a determinant of policy (Cooper, 1987; Heeley, 1989). Equity and social goals, however, have been of less significance and Henry (1994) argues that leisure policy generally in the United Kingdom has not dealt with the issue of the leisure-poor and uneven access to tourism and leisure facilities.

The expansion of tourism policy at the local level in the United Kingdom has involved the rapid growth of locally based urban and rural tourism initiatives (Law, 1993; Shaw and Williams, 1994). Explanations for the increasing role of the local state agencies in the process of tourism promotion, image reconstruction and place marketing have linked policy to economic restructuring and the new forms of urban entrepreneurialism that have emerged in a post-industrial society (Harvey, 1989). Ashworth and Voogd (1990) argue that economic restructuring and the growth of a professional service class stimulate the emergence of new local economic development responses based on marketing and tourism in order to diversify the local economy. Similarly, Urry (1990) argues that economic stimuli to local tourism policy development include deindustrialisation, market differentiation, and the fragmentation of ownership in tourism. In addition, local conservation groups, who are often dominated by members of the new service class, have been an important influence on local tourism policy (Urry, 1990). Policy initiatives, therefore, seek to link economic development to the structural determinants of tourism consumption which are social and cultural as well as economic. Urry (1993) identifies key changes in the types of European tourism in the 1990s including increases in globally responsible tourism, a rise in second homes/time shares, increased tourism in peripheral and rural areas, the growth of tourism among young and old, the expansion of city-centre tourism and the development of heritage and cultural tourism. The last of these changes relates to the continued growth of what Poon (1989) regards as the key growth areas in a 'new tourism' typified by segmentation and flexibility in terms of product compared to mass packaged tourism.

The economic and social construction of tourism policy is, of course, mitigated by a range of political influences on policy evolution. As with local economic policy generally, relations with central and other tiers of government have been an important determinant of local tourism policy throughout the United Kingdom (Bull and Church, 1995). Indeed, the restrictions placed on local authority economic policy in the 1980s were partly responsible for the growth of tourism initiatives as local agencies saw this as an area where they could develop new initiatives and seek funding from other agencies such as the Tourism

Development Action Plans supported by the English Tourist Board (Urry, 1990). In addition, the European Union has since the late 1970s provided funds for a variety of tourism initiatives that have often involved local agencies (Pearce, 1988; Sinclair and Page, 1993).

A further broad political influence on tourism policy at the local level has been concern for the environment. Dowling (1992) notes how the issue of the tourism–environment relationship has been under discussion by international agencies since the 1950s, but it was only in the mid-1980s that this relationship began to be analysed and understood. In some resorts, however, local government has been dealing with the environmental problems created by overuse for many years. The English Tourist Board (1991) has established guidelines for managing the tourism–environment relationship. These suggest that the key aims for local government should be improving management information, better coordination and additional resourcing, combined with practical management strategies to assess capacity, manage tourism transport, provide marketing information, conserve key sites, control development and involve the local community (Shaw and Williams, 1992). A few local authorities in cityport regions have gone much further and developed detailed coastal zone management strategies (May and Heeps, 1994). At the local level, however, there is marked variation in the political and management approach to the tourism–environment relationship ranging from 'a complete laissez-faire attitude to open hostility, as well as encompassing a period of strong encouragement to gain economic benefits from tourism' (Shaw and Williams, 1992, 53). Another political influence on recent tourism strategies, especially those relating to culture, has been the experiences of other countries in Europe and America (Lim, 1993).

These differing tourism policies all have problematic economic, social and political impacts (Ashworth and Voogd, 1990). Political difficulties arise since without the coordination of the image promotion strategies of neighbouring locations unproductive competition may occur (Urry, 1990). The economic effect of many tourist and image promotion policies has been questioned, partly due to the lack of information on the impacts of such policies (Pearce, 1992; Paddison, 1993). Also, direct job creation in tourist facilities is often limited and what does occur is low-paid and part-time (Law, 1992; Townsend, 1992). In addition, negative social and cultural effects arise as a result of the subversion of genuine cultural identities in the pursuit of new images (Boyle and Hughes, 1991). The use of heritage and cultural facilities are socially biased towards upper and middle income social groups (Lim, 1993; Prentice, 1994), which may promote alienation among groups whose needs and interests are less directly served by new tourism attractions.

It is clear, therefore, that an understanding of state policies towards tourism in cityport regions will require a broad interpretation that takes account not only of the more predictable economic and political determinants of policy, but also the social and cultural influences that play a crucial role in shaping the nature of contemporary tourism. The remainder of this chapter focuses on the evolution of tourism policy in three cityport regions in the UK and aims to illustrate how initiatives that were primarily developed as economic policies have an important related social and cultural dimension. In two of the case studies, Dover and Teesside/Hartlepool, the policy measures under consideration were developed in the late 1980s and early 1990s and hence the impacts are only just starting to emerge. For this reason the consequences of policy are considered in less detail.

TOURISM IN UNITED KINGDOM CITYPORT REGIONS – GENERAL ISSUES

Discussion of the case studies is based on a broad view of the measures that constitute tourism policy and the initiatives analysed are those that the agencies in each area define as being part of their tourist strategy. This approach is necessary due to the long-running and problematic debate over what constitutes tourism (see Gilbert, 1990; Williams and Shaw, 1991a). From a technical viewpoint tourism in the United Kingdom is understood to involve an overnight stay away from home of 24 hours or more, but the notion of a 'tourist gaze' involves many activities and social practices that do not require such a stay away. As Gilbert (1990) notes, 'the phenomena of tourism would seem to comprise a range of products formed from the segmented portions of the larger industries'. A number of the policy initiatives discussed below are concerned more with leisure than tourism, but are included in the case studies since they are defined by local agencies as being part of their tourism strategies.

The case studies of Dover in south-east England, Teesside/Hartlepool in north-east England, and the London Docklands have in part been chosen due to their differing local economic, political and social contexts. A comparison of the tourism strategies adopted in these varied localities illustrates a range of policy approaches and implementation methods. In each case study the recent evolution of policy is discussed in terms of its economic aims and the agency's perceptions of the future production and consumption of tourism in the local area. The role of cultural and social factors in determining policy is also considered and is followed by a discussion of the different policy delivery mechanisms and outcomes.

The agencies with responsibility for tourism vary slightly between the three cityport case studies. Due to the fragmented nature of tourism in Britain, a series of national and regional agencies exist to promote tourist locations and the industry as a whole. At the national level the British Tourist Authority promotes tourism abroad, while domestic tourism is the concern of the English Tourist Board and a series of regional boards. In the case study areas the relevant regional boards are the London Tourist Board, the South East Tourist Board and the Northumbria Tourist Board. In each area, however, recent attempts in former port areas to stimulate the tourist industry and change local images have been mainly initiated by more localised agencies.

County and local councils in the United Kingdom both have a responsibility for tourism under a series of government acts. The Planning Policy Guidance Note 21 (Department of the Environment, 1993) indicated that central government identified a clear role for local authorities to promote and manage tourism through local plans, especially in tourism development. Nevertheless, the number of authorities with tourism officers has grown substantially and many former industrial locations that had images not traditionally associated with tourism have started to develop strategies despite negative perceptions of their areas (Buckley and Witt, 1985, 1989). Indeed, the three cityport case studies were also locations that in the past did not attract large numbers of tourists, but Dover certainly saw many passing through the area due to its passenger port function.

In two of the case study areas, London Docklands and Teesside/Hartlepool, much of the impetus for tourism policy has come from locally based, but centrally controlled, Urban Development Corporations (UDCs). The UDCs were the main element of central government policy to regenerate depressed and derelict urban areas (Imrie and Thomas, 1993). The London Docklands Development Corporation (LDDC) was established in 1981 and the Teesside Development Corporation (TDC) in 1988.

THE DOVER CITYPORT REGION – ECONOMIC DIVERSIFICATION AND A NEW CITYPORT REGION IMAGE

Dover in the south-east corner of England is Europe's busiest ferry port (Dover District Council, 1994). It is located in the county of Kent in close proximity to other lesser Channel ports such as Folkestone and Ramsgate. In terms of population it is not a large cityport region compared to those in the United Kingdom that are part of conurbations. Dover local authority district (see Figure 11.2) includes the town of Dover and its hinterland with a population in 1992 of 99 400 (Kent County Council, 1993). The port performed well in the 1980s with passenger traffic increasing by 35% between 1981 and 1989 and freight traffic by 50% (Dover Harbour Board). The local economy that in 1991 contained 43 000 jobs is very dependent on the port. One estimate suggested that between 14 and 18% of local employment in the town of Dover is directly related to the port, with many other jobs dependent on multiplier effects (Dover District Council/ Eurotunnel, 1987). Nevertheless, the local economy performed favourably in the late 1980s as unemployment fell from 11.4% in March 1986 to 5.1% in May 1990, just below the national average of 5.7% (PACEC, 1991). The recession of the late 1980s has seen unemployment rise to just over 11%. Politically the area in the 1980s was represented by Conservative members of Parliament and a Conservative local council, but in 1993 no single party was in overall control of the council. Part of the port has closed in recent years and the old Wellington Dock is now being marketed as a location for mixed tourism development based upon the waterfront. In many other cityport regions such waterfront redevelopment tends to be the focus of tourism policy, whereas in Dover the waterfront site is merely a small part of a much wider tourism strategy.

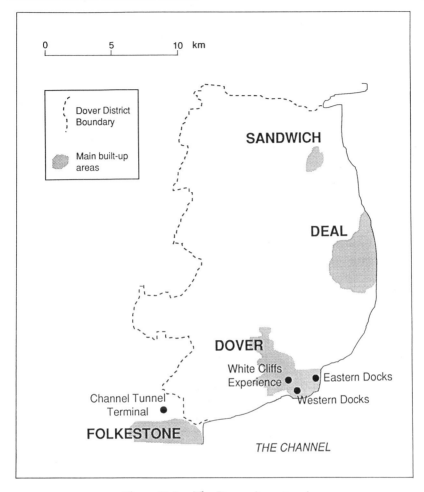

Figure 11.2. The Dover cityport region

The local council has now developed a major tourism strategy involving considerable public expenditure. The increasing emphasis on tourism as one of the key elements of the local economic development strategy started to emerge in the mid-1980s. In part this was a result of the early 1980s recession, but the key economic stimulant of this policy development was the potential threat of the Channel Tunnel and the Single European Market. Competition for passenger and freight ferries from the Channel Tunnel and the reduced needs for Customs facilities as a result of the Single European Market were perceived as leading to job losses in the port and port-related industries. One early forecast suggested that 4000 jobs in Dover could be lost as a result of the Channel Tunnel (Channel Tunnel Joint Consultative Committee, 1987), but more recent studies argue that this is an overestimate due to the competitive strategies adopted by the ferry companies (Kent Ferry Operators, 1990).

Nevertheless, the threat of employment decline encouraged the development of economic diversification strategies for the wider cityport region (Dover District Council, 1986). Certain political factors were encouraging the development of this policy since the tourism sector had already been identified as a potential growth sector by Kent County Council and national government were willing to provide money for the maintenance of Dover Castle, one of the town's chief attractions (Dover District Council, 1987). An initial policy development was the establishment of the East Kent Initiative, a joint attempt by six local councils and the county council to develop a promotional strategy for the whole of East Kent aimed at tourists and investors (East Kent TDAP, 1990). Dover played a role in the East Kent Initiative but was concurrently developing its own policies which gained a considerably higher profile than the joint initiative.

A series of studies and visitor surveys of existing tourism facilities in the 1980s identified the under-performance of certain attractions, such as the castle, and a local hotel catering sector that viewed future growth prospects as limited (Peat Marwick McLintock, 1988). Consequently the company responsible for the Channel Tunnel adopted a negative view regarding the future of tourism in Dover (Eurotunnel, 1986). On a more positive point it was noted that a very large number of tourists passed through the area and if a small proportion could be persuaded to stay for a short period then the tourist sector would benefit. Short-stay tourism was a major tourist growth sector in the 1980s (Williams and Shaw, 1991). It was this market sector, along with day visitors, that Dover chose to target. It was hoped that some of the short-stay and day visitors would include tourists from mainland Europe *en route* to other locations in the United Kingdom and the most optimistic forecasts suggested that in future the expenditure of 600 000 day visitors and 300 000 overnight visitors could create 1400 jobs in tourism (Peat Marwick McLintock, 1988).

The negative image of Dover as a port and as an area that was simply passed through by tourists was perceived as a major obstacle to increasing tourist activity and expenditure (Dover District Council, 1986). Changing the historical and cultural image of Dover was at the centre of the policies to be developed, but the promotional initiatives connected with a new image were to be strategic in approach. The aim was to develop an image that would relate to the whole cityport region of Dover and its hinterland and thus spread the benefits of any tourism growth throughout the cityport region. The Dover tourism strategy was agreed in 1987 (Dover District Council, 1987). Over 40 significant attractions exist within the district (Dover District Council, 1994) and a marketing and image reconstruction exercise was designed to promote them as a whole. In addition, the series of studies in the mid-1980s argued that the marketing strategies should be aimed at harnessing the economic benefits of the cultural trends that were stimulating the growth of the heritage industry (Peak Marwick McLintock, 1988).

The image reconstruction strategy developed since 1988 involved renaming Dover and its hinterland as the White Cliffs Country. This is a common approach adopted elsewhere which seeks to give space a coherent identity, readily accessible through the existing cultural images and historical knowledge that tourists already possess. Prentice (1994) identifies in England and Wales 28 tourist 'countries' (e.g. Shakespeare Country), one tourist 'city' (Beatle City Liverpool) and one 'kingdom'. The successful tourist 'country' locations are usually based on several major attractions (Prentice, 1994). Dover recognised that its cliffs, downland, castles, ancient monuments and hinterland towns represented significant attractions, but consultants employed by the council argued that many of these facilities needed improving and a clear tourist identity required a major new attraction to 'bind' the others successfully together (Peat Marwick McLintock, 1988). At the time the operational port itself was a tourist attraction often being viewed from above from the top of the White Cliffs. The port and castle were the most visited attractions in Dover, but were not considered sufficient to attract large numbers of short-stay visitors who would visit other attractions elsewhere in the cityport region. Furthermore, consultants argued that the private sector was not interested in investing in a major new facility (Peat Marwick McLintock, 1988).

Figure 11.3. The Norman Crescent in the White Cliffs Experience, Dover, from which visitors can see the Norman remains of the church of St Martin-le-Grand. (Photograph courtesy of White Cliffs Experience, Dover)

Figure 11.4. The White Cliffs Experience, Dover. Waxworks form part of the presentation of Dover as a frontier town, with ancient Celts preparing to face invading Romans. (Courtesy of White Cliffs Experience, Dover)

A significant new attraction in the centre of Dover was established in 1991 by Dover Council, using £14 million of public money and known as the White Cliffs Experience. Designed by John Sutherland (the designer of the Jorvik Centre in York), the White Cliffs Experience seeks to integrate Dover's Roman remains with an historical account of its role as a frontier town (Piotrowski and Page, 1991). Visitors move through the building viewing re-creations of key events in Dover's history (ancient Britons preparing to fight Roman invaders, early channel crossings, and the Battle of Britain in the Second World War; Figures 11.3 and 11.4) and are encouraged to visit other attractions not only in Dover but also in the surrounding area. This type of municipal enterprise investment – managed by Heritage Projects PLC with the local council maintaining an equity share – is more normally associated with socialist councils in Britain in the 1980s, but at Dover the Conservative council took the view that only a major public investment would achieve the scale of project required. It is ironic that this municipal enterprise was developed at the same time as the Conservative central government was passing legislation in the form of the 1989 Housing and Planning Act to discourage local authorities from taking this type of approach (Campbell, 1990).

Measuring the economic effects of such initiatives is fraught with difficulties. Piotrowski and Page (1991) warned that there are some significant risks associated with the White Cliffs Experience stemming from the nature of the visitor experience, the limited growth in domestic UK tourism and the potential threats to school excursion budgets. The local council, however, points proudly to visitor numbers: 235 000 people visited the White Cliffs Experience in its first year of operation compared to a predicted target of 150 000 (Dover District Council, 1993). Before this, annual visitor numbers to Dover Castle had never exceeded 100 000. Visitor numbers to the White Cliffs Experience fell slightly to 230 000 for the year 1992/3 but still generated an additional expenditure in Dover of between £1.5 million and £2 million per year (Dover District Council, 1993). Less satisfactorily, the White Cliffs Experience does not seem to attract large numbers of short-stay visitors. Just under 70% of visitors are from the surrounding county of Kent itself.

The most recent Tourism Business Plan for Dover estimates that total visitor numbers to all attractions in White Cliffs Country are now 650 000 per annum, but that a fresh marketing campaign is required to increase the number that stay overnight (Dover District Council, 1994). The dominance of tourism in the council's approach to economic diversification means that marketing and promotion is now a clear priority for local economic policy. One-third of the economic development budget for 1994/95 was earmarked for marketing compared to under 10% on enterprise support (Dover District Council, 1993). In addition, a new product is being developed: the Historic Town Trail which, in keeping with the strategic aim of policy, seeks to encourage visitors to explore more of Dover's hinterland. Clearly the White Cliffs Country tourism strategy has generated visitors and increased tourist expenditure. This may be reflected in the fact that jobs in the hospitality sector (hotel and catering) in Dover grew by 2.3% between 1989 and 1991 compared to a decline in the South East region of 2.5% (Census of Employment, 1989, 1991). Whether the long-term economic effects will represent a good return on the major public investment remains to be seen.

The cultural and social consequences of this initiative are significant. The White Cliffs Experience promotional literature continually stresses its authenticity, that it is 'academically researched' and the attached museum is 'a place for research at all levels'. Clearly, this is designed to satisfy the desire for 'authentic' attractions among certain visitors that Thomas (1989) has identified elsewhere. The symbolic role of the White Cliffs Country is of interest since it seeks to relate visitor experiences to a nationalistic social identity. Dover is renamed 'Britain's front-line story' and many of the scenes in the White Cliffs Experience relate to Britain's military history, especially the Second World War. Urry (1990) notes that the design of 'extraordinary' tourist gazes will involve creating imagined communities to which tourists can relate. As Anderson (1983) and Urry (1993) argue, it is imagined communities that form the basis for social identities since they allow individuals to integrate their interpretations of time, space and memory. Many heritage-based tourist attractions, however, seek to present an imagined community based on a local or regional history in order to give the tourist experience a distinctive nature. In the case of Dover the construction of the objects that form the tourist gaze attempts to integrate two imagined entities, namely Dover as an historical frontier town and the complex sense of belonging that is the basis of nationalism. This may seem an innovative policy since the use of nationalistic tropes may add symbolic value to the locally based reconstruction of Dover's historical heritage. In this way the tourist may experience a greater sense of belonging to the objects that are presented in the White Cliffs Experience. Alternatively, the incorporation of nationalist symbols, especially military ones, in the tourist environment may negatively affect certain key social groups. Visitors from Britain will no doubt have mixed reactions depending on their social characteristics to a nationalism that ranges from tattooed ancient Britons to World War Two Spitfire aeroplanes. The responses of non-domestic visitors will be harder to predict but the nationalistic basis to many attractions elsewhere in Britain does not harm visitor numbers. The overseas visitors are of course a key element in the hoped-for increase in overnight stays.

From a policy point of view the approach to economic diversification in Dover based on tourism has indicated the possibility of strategic action that attempts to spread the economic benefits of tourism throughout the cityport region. The strategic focus was further enhanced by the development in 1987 of the Transfrontier Development Programme, a cooperative venture between Kent County Council and Nord-Pas-de-Calais, France. This has successfully obtained funding from the European Union Interreg programme for border regions, a large proportion of which has been used to support a number of tourism initiatives in the coastal areas on either side of the Channel (Sinclair and Page, 1993; Church and Reid, 1994). The next decade will reveal whether these tourism initiatives can compensate for any job losses that may occur from port restructuring in response to Channel Tunnel competition.

TEESSIDE/HARTLEPOOL – ECONOMIC DIVERSIFICATION, URBAN REGENERATION AND FLAGSHIP SITES

The County of Cleveland in north-east England includes the towns Middlesbrough, Stockton and Hartlepool that collectively make up the conurbation of Teesside (Figure 11.5). This is one of England's major cityport regions, with a number of port facilities based along the estuary of the River Tees, and an industrial history based upon traditional manufacturing industries

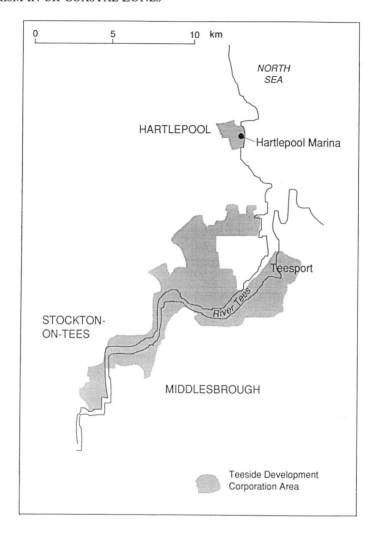

Figure 11.5. Teeside/Hartlepool

such as steel, chemicals, shipbuilding and heavy engineering. In 1993 there were 200 000 jobs in Cleveland of which 28% were in manufacturing. Economic restructuring and the privatisation of key industries such as steel led to marked job losses at certain points in the last 20 years. Consequently, employment in the manufacturing sector fell by 25% in the 1980s (Cleveland County Council, 1993a). In August 1993 unemployment in Cleveland stood at 16.8% compared to 10.4% nationally (Cleveland County Council, 1993a).

Teesport remains, however, England's 6th busiest non-oil port (Tees and Hartlepool Port Authority, 1993), but certain parts of the port have closed in the last 20 years. One of these is the South Docks at Hartlepool which forms the focus of this case study. The town of Hartlepool itself has experienced marked reductions in steel employment in the last 20 years which helped to contribute to the high rate of unemployment in 1993 of 15.7% (Cleveland County Council, 1993a).

The need for diversification of the local economy in Teesside generally, and Hartlepool in particular, has been apparent since the 1950s and there have been various central government regional policies charged with this goal (Hudson, 1989). In the mid-1980s tourism became part of the economic development strategies of a number of local agencies. Primarily these policies were designed to generate jobs in areas of high unemployment and provide the local economy with a more diversified economic base (Cleveland County Council, 1993b). The last few years have seen attempts to link the policies of the different agencies in the city port region as part of a more strategic approach to tourism.

The policies developed were also a response to the existing patterns of tourism. Visitor surveys in the late 1980s indicated the limited development of tourism in many parts of the county (Cleveland County Council, 1989). In the south of the county certain locations attracted large numbers of visitors. One is another example of a tourist 'country', Captain Cook Country, which has become a successful heritage attraction. Historic inland towns such as Guisborough and attractive coastal areas outside the Teesside conurbation are popular destinations for day visitors. In addition, a more recent survey indicated the importance of business tourism in supporting the hospitality industry in urban areas (Cleveland County Council, 1993c). The other major group of overnight visitors, besides business tourists, were individuals visiting friends or relatives (Cleveland County Council, 1989). The locations that attracted visitors to Cleveland were primarily in the countryside surrounding the Teesside conurbation, some of which is part of the North Yorkshire Moors National Park (Cleveland County Council, 1989).

In this economic context the county council, in conjunction with the Northumbria Tourist Board, developed a series of Tourism Strategy Action Programmes designed to attract more visitors to Cleveland generally and in particular to develop urban tourism locations (Cleveland County Council, 1993b). The urban initiatives are aimed at day tourists from the large northern conurbations, people visiting friends and relatives in Cleveland, and the business tourist. The overnight-stay market that Dover targeted is not such a priority aim, in keeping with the aims of the local Hartlepool Borough Council (Hartlepool Borough Council, 1994). Hartlepool has a problematic past image and is not normally associated with tourism although it does have a number of historic and former port locations that might attract additional visitors (Hartlepool Borough Council, 1993). In addition, industrial tourism in the form of the Energy Information Centre at Hartlepool Nuclear Power Station attracts about 30 000 visitors a year (Hartlepool Borough Council, 1994).

The development of tourism plans by the county council, the regional tourist board and the local council were also shaped by relations with central government which in 1988 established the Teesside Development Corporation (TDC), charged with the physical, economic and social regeneration of 12 000 acres of land in the local area. The regeneration strategies of Urban Development Corporations (UDCs) focus on the use of public money for land clearance and infrastructure development with the aim of levering in significant private investment attracted by incentives and limited planning bureaucracy (Imrie and Thomas, 1993). The Conservative ideological basis to TDC policy has led to some disagreements with certain local Labour councils in Teesside (Cleveland County Council, 1993d). Generally, however, the local councils have cooperated and local council leaders have seats on the Board of the TDC. Some commentators, however, have argued that the new alliances between the TDC, local councils and local business interests have led to the exclusion of certain groups in the local community from the policy process (Robinson and Shaw, 1991) and that local employment creation estimates have been somewhat exaggerated (Robinson *et al.*, 1994).

An initial background paper completed for central government prior to the establishment of the TDC argued that despite the public image of the area as a depressed industrial region there was scope for the development of major tourism and leisure facilities in the TDC area. The Hartlepool South Docks were identified as a potential flagship site for tourist development (Coopers and Lybrand, 1987). Since then the TDC has spent £80 million developing the

Figure 11.6. Hartlepool: new yachting berths and new housing. (A. Church)

Hartlepool Docks as a marina with adjoining housing development and tourist attractions (Figure 11.6). The demand for marina berths was high and the second phase of the development was quickly undertaken so that there are now 165 yacht berths with more under construction (TDC, 1993). An historic ships heritage centre has been established and prospects for the re-use of the docks as a tourist attraction were increased considerably when in 1993 it appeared that the TDC had persuaded the Imperial War Museum in London to choose Hartlepool as the location for the Northern Outstation of its historic ships collection. This facility was predicted to attract 400 000 visitors per year (TDC, 1993) but in July 1995 the Imperial War Museum withdrew from the project. Over 220 houses, however, have been constructed on sites adjacent to the Docks, about a third of which are low-rent social housing units.

In comparison to tourism developments at Dover the Hartlepool site contains the features more typically associated with tourism waterfront developments in cityport regions (Hoyle and Pinder, 1992). Public money was spent on land reclamation and the infrastructure required for the marina development which is now run by a private sector company, Yacht Haven Management Ltd. Consequently, the increased amenity value of the area has been used to market adjoining sites for private sector housing development and for hotel and catering outlets.

The economic and social consequences of marina and tourism-based waterfront developments have been examined in some detail in other locations (Law, 1988, 1993). The direct job-creation effects of the marina in Hartlepool are limited (Cleveland County Council, 1993d), but the TDC points to the 300 000 visitors to the development in 1993 partly as a result of the Round Britain and Ireland Yacht Race (TDC, 1993). In fact in the wider district of Hartlepool the hotel and catering sector lost over 300 full-time equivalent jobs between 1981 and 1991 (Census of Employment, 1981, 1991). The UDC argues that the construction of new restaurants, pubs and hotels in the docks area in the near future will generate employment (TDC, 1993). In the northern region as a whole, however, Thomas (1995) argues that the employment impacts of tourism are often exaggerated by local agencies.

The social and cultural effects of such waterfront locations have also been discussed with reference to other locations (Boyer, 1992). Indeed, Harvey (1988) asks how many more such environments 'can we stand'. The redevelopment in Hartlepool is promoted by the TDC as the 'Hartlepool Renaissance' and the remaking of this former port space to create a 'deluxe' consumption environment is mainly designed, despite certain large-scale public events, to promote private, individualistic consumption. The symbolic marina environment at Hartlepool provides powerful signals to encourage individual participation in the consumption of private goods. Interestingly, however, the waterfront area is still primarily used by local Hartlepool residents for informal leisure activities such as dog walking (Cleveland County Council, 1994). This may indicate that a coherent tourist gaze has yet to emerge in Hartlepool, and to what extent the future public spaces will accommodate informal activities alongside the tourists remains to be seen. The concern of other local agencies is whether the visitors attracted to Hartlepool will in the long term produce economic benefits felt over the wider cityport region. In Liverpool the Merseyside Tourist Board argues that the key tourist sites on the waterfront, such as the Albert Dock, have been crucial elements in the emergence of a broader tourist-related economy that now employs 14 000 people throughout the cityport region (*Financial Times*, 2 July 1992). Indeed, the county of Merseyside experienced a 12.3% rise in hospitality employment between 1989 and 1991 which was the largest increase for any of the 67 counties in Great Britain (Bull and Church, 1994). In the Teesside cityport region the county council, the regional tourist board and local councils have devised tourism policies to encourage the development of the industry throughout the region. Such policies have not been pursued with the same level of resources as in Dover where the White Cliffs Country has been advertised regularly on national television. This is of course partly a reflection of the many pressing priorities facing local agencies in Teesside. At the moment it is the flagship sites of the TDC such as Hartlepool and two other sites in Middlesbrough that have been the focus for large-scale tourism and leisure investment. In such a situation it is the policy aims of central government urban regeneration that are a key determinant of locally based tourism strategies for the cityport region.

LONDON DOCKLANDS – MARKET-LED PLANNING AND TOURISM POLICY

London Docklands (Figure 11.7) is not strictly a cityport region. The area for which the London Docklands Development Corporation (LDDC) is responsible contains the former docks in an area of London that now contains very few port activities. The nature of the LDDC's market-led regeneration strategy and its problematic consequences have been extensively debated (see Brownill, 1992). Given the extensive LDDC expenditure on the marketing of London Docklands and a constant concern with image promotion, it is perhaps surprising to find that although the LDDC was established in 1981, for a variety of reasons a tourism and visitor strategy only emerged in 1988–89.

The initial aims of the LDDC suggested that tourism would be a significant concern. In 1983 the LDDC defined a series of economic growth sectors which it hoped to expand in Docklands and tourism was one of these (LDDC, 1983). This was hardly surprising since Docklands was already home to a number of successful tourism initiatives close to the City of London. HMS *Belfast* is located at the western end of Docklands near Tower Bridge. St Katherine's Dock is an important tourist attraction developed in the 1970s based around a marina, a pub, a hotel and a few speciality shops. In subsequent years major office developments were also completed in this location (Ledgerwood, 1987). Indeed, the LDDC was not the first organisation to identify the potential of tourism in

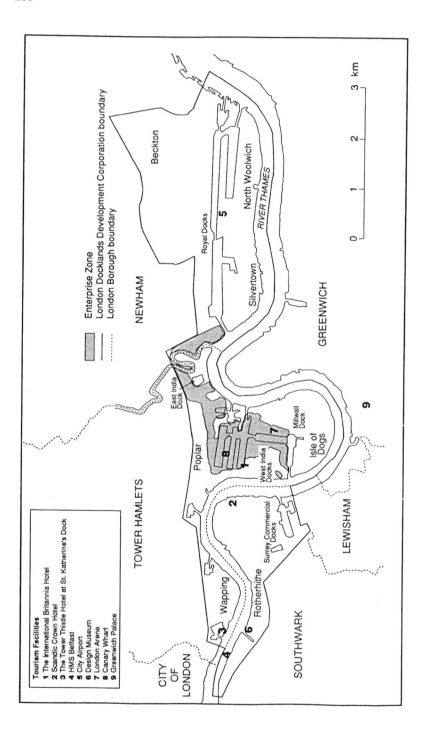

Figure 11.7. London Docklands

London Docklands. An earlier regeneration agency, the London Docklands Study Team, identified in 1973 five possible redevelopment strategies for the area that contained proposals for golf courses, marinas, leisure centres, maritime museums, camp sites, hotels and equestrian centres (London Docklands Study Team, 1973). In 1979 a feasibility study for holding the 1988 Olympics in London examined the potential of Docklands as a base for the games and argued that the construction of new facilities in Docklands would have future tourism potential (GLC, 1979).

Under the LDDC in the early and mid-1980s, a number of tourism initiatives were proposed, some of which were never completed. Perhaps the most significant of these was for a location called North Quay on the Isle of Dogs. This waterfront site contained some of the oldest dock warehouses in Docklands dating from the early 19th century. Initial proposals suggested that this could be a 'people draw' area, modelled on Quincey market in Boston, containing speciality shopping, a hotel and a heritage function (LDDC, 1984). The Museum of London has a major collection of Docklands artefacts and was interested in developing a Docklands Museum on part of this site. The 'people draw' scheme faced a number of problems. Private developers were unsure that enough visitors could be attracted to the site and were clear that the neighbouring public housing would be unable to generate the demand for speciality shops. Building conservation legislation also created numerous obstacles to finding a use for the old warehouses since low ceilings could not be changed and iron window bars could not be removed. This scheme was soon discarded when in 1986 proposals were put forward for the massive office development at neighbouring Canary Wharf which has become a 'people draw' area in its own right.

The Docklands Museum continued its search for a location, but the LDDC under the direction of central government was not willing to invest large amounts of public money in the scheme and argued that a private backer should be found. A private sector sponsor did not emerge and the Docklands Museum artefacts remained in storage elsewhere in London. This issue may finally be resolved in 1995 with a museum established at West India Quay near Canary Wharf. The initial insistence on finding private support for the museum was of course very different to the approach adopted in Hartlepool where public expenditure was used to develop a maritime heritage centre.

Other developments based on providing speciality shopping and leisure facilities for tourists and local office workers were constructed. The Hays Galleria development is reasonably successful mainly due to its location close to major City of London office developments. Elsewhere in Docklands an old skin warehouse was converted into a speciality shopping precinct named Tobacco Dock which contained two imitation historic ships to create a tourist attraction. In the early 1990s this development went into receivership partly due to its inaccessible location and the lack of nearby office developments. The current owners are redesigning it as a 'factory shopping' centre which places less emphasis on its role as a tourist attraction.

Throughout the mid-1980s the lack of a stated tourism strategy was partly the result of the fact that commercial office development and housing construction were proceeding rapidly after 1985. Tourism facilities, therefore, were seen as something that would follow in the wake of office development. One former LDDC officer claimed that tourism was perceived within the LDDC as 'the icing on the cake' of large-scale office development. In 1988, however, a consultant's report formed the basis of a tourism strategy for Docklands. The report proposed a three-level strategy based on major developments along a River Thames corridor, on some significant attractions near the docks and on a number of more isolated historic attractions spread throughout the area (Llewelyn-Davies, 1987; Beioley et al., 1988). The attractions would be linked up by the new Docklands Light Railway and a commuter Riverbus, both of which started to operate in the mid-1980s. These two infrastructural developments were seen to have considerable tourism potential (Beioley et al., 1988). Ironically, it was the downturn in the commercial office market in the late 1980s that provided a further stimulus for the development of an LDDC tourism strategy. In order to revive the commercial image of Docklands a new marketing team was established in 1989. As part of this initiative a Docklands tourism marketing strategy was developed (LDDC, 1989).

In the early 1990s, however, the LDDC was rationalised to focus its activities on infrastructure policy and to prepare for its eventual winding down. Tourism was not considered a key priority in this process and a number of staff concerned with tourism were given other responsibilities. The current tourism strategy, therefore, is perhaps less extensive than was imagined in the late 1980s. The development of a riverside tourism corridor became more problematic when the commuter Riverbus went bankrupt and ceased to operate. Indeed, recent promotional material guides tourists towards Canary Wharf (Figure 11.8), a walking heritage trail and certain key sites, such as the Design Museum (Figure 11.9) and the National Maritime Museum at Greenwich just outside Docklands (LDDC, 1993, 1995). The LDDC continues to provide a Visitor Centre and plenty of promotional literature. An officer is responsible for implementing a visitor strategy and the property team in 1994 started to promote seven vacant sites as possible locations for hotels (LDDC, 1994).

Nevertheless, the construction of a series of new tourist attractions combined with those that existed prior to the LDDC means that Docklands now has a network of tourist facilities ranging from the new national Design Museum at the western end of Docklands to an artificial ski slope at the eastern edge of the area. The Docklands Light Railway is a tourist attraction in its own right and LDDC expenditure has transformed a number of run-down historical buildings into tourist attractions. Since 1981 major hotels have opened in Docklands, mainly designed to serve the business market, the Docklands Arena provides a new indoor sports and entertainment facility, and four watersports centres now exist in the area. In 1991 tourism employed over 5000 people in a Docklands economy of 54 000 jobs (Insight Social Research, 1991). This development of tourism,

Figure 11.8. London Docklands: the Scandic Crown Hotel with Canary Wharf in the background on the opposite bank of the River Thames. (Courtesy of London Docklands Development Corporation)

however, is closely linked to the expansion of other sectors of the Docklands economy. The area receives 1.2 million visitors a year, a significant proportion of whom are business tourists (LDDC, 1994).

The changing emphasis over time on tourism policy in Docklands is perhaps expected given that the LDDC's regeneration strategy has sought to adjust to changes in market forces. The question that remains, however, is to what extent the LDDC's policy is one element in a chain of events that has resulted in a missed strategic opportunity for tourism in London. Page and Sinclair (1989) note that the development of facilities in Docklands has significant implications for the tourism sector in London as a whole. The problematic geographical concentration of hotels in central London has been accompanied by poor standards, high prices and a shortage of budget accommodation (Page and Sinclair, 1989). Tackling such important strategic issues in the capital is not easy given the wide variety of agencies with differing agendas involved with tourism (Bull and Church, 1995). Indeed, Page (1988) argues that cooperation between the LDDC and tourism agencies is crucial to the success of London's tourism industry. Hotel development in Docklands certainly had the potential to spread tourism accommodation and activity away from central London (Page and Sinclair, 1989). In 1989 development proposals existed for 11 new hotels in Docklands with a combined capacity of over 4000 beds (Page and Sinclair, 1989).

Figure 11.9. London Docklands: the Design Museum and new upper-income housing developments. (Courtesy of London Docklands Development Corporation)

Many of these projects were speculative and never came to fruition due to the national recession. The two new hotels in Docklands added just over 800 rooms to the London total and a new youth hostel provides 70 much needed budget accommodation rooms.

The impression still remains however, that the development in the early 1980s of a more strategic approach to hotels and tourism in Docklands might have had a greater impact on the pressing London-wide problem of visitor accommodation. The LDDC, of course, does not bear sole responsibility for the process of policy evolution. Since the Greater London Council was abolished in 1986, the London Tourist Board took the lead in developing London-wide tourism planning. Despite the production of innovative Tourism Strategies for London (London Tourist Board, 1987, 1993) tourism policy in the capital, especially as regards land-use issues, lacks a strategic overview (Bull and Church, 1995). The London Planning Advisory Committee has recently endeavoured to tackle the need for a broad approach to hotel provision in London (LPAC, 1994). It is possible that the LDDC's current marketing of seven sites for hotel development and a further site for a major new conference centre may in future result in the redevelopment of Docklands, alleviating some of the tourism problems of central London. Recent events suggest, however, that this is not a priority for central government. In 1994 the LDDC was part of a public/private partnership coordinated by the London Tourist Board that tried to adopt a more strategic approach to tourism in East London. This led to the development of an East London Tourism Initiative which was submitted to central government's Single Regeneration Budget, but failed to receive funding.

A strategic approach to hotel development in Docklands might well have attracted considerable opposition locally since it would appear as a policy designed for the benefit of the business élite. On many occasions the national or regional interest has been used as a justification for a locally unpopular policy in Docklands. Tourism provision has, however, had an important role in relation to the local community. The LDDC's annual reports over the last five years have all discussed tourism and leisure alongside the local community. Images are provided of local residents using new leisure facilities. Presented in this manner tourism and leisure facilities seem to play a symbolic role legitimating state policies towards Docklands, as they have often done elsewhere (Shaw and Williams, 1994). The on-going problematic reality of Docklands that includes high unemployment and poor housing is ignored in favour of this alternative leisure-based image. Textual analyses of Docklands have discussed the symbolic meaning of the new built environment and have tended to stress its role as an expression of the dominance of financial power and new service class residents in Docklands (Short, 1989; Crilley, 1993). Alongside the massive tower at the heart of Canary Wharf the new tourist and leisure gaze may seem quite innocuous, but its social and cultural role in the legitimation of mass urban regeneration may be just as important.

CONCLUSIONS

The three case studies discussed in this chapter illustrate that somewhat different approaches have been adopted to tourism in cityport regions and it is perhaps misplaced to suggest that such policies are predictably similar. The uneven development of tourism policies does indicate that currently there are certain key choices facing local agencies. The general trends in tourist consumption are important economic determinants of tourism policy, but each authority has developed initiatives targeted on certain market segments. Law (1993) outlines the various market segments that could be the targets of urban tourism. In London Docklands business visitors have become a key target sector for tourism development. In Hartlepool day visitors and people visiting friends and relatives were the initial aim of strategy, but the overnight visitor may become a future target. Overnight visitors were always a key element in Dover's strategy, but this may require new initiatives as, currently, day visitors dominate. In addition, there is a certain amount of local discretion over the degree to which tourism policy is strategic in its aims. A region-wide strategy has been developed in Dover; in Teesside/Hartlepool tourism strategy is focused on certain flagship sites and in

London Docklands a local strategy has been initiated but it may be that an opportunity to develop an important strategic policy for London but has not been taken. Local agencies also face key choices over the nature of cultural and historical image reconstruction and the way space is redefined in cityport regions. The Hartlepool initiative has many similarities to port redevelopments elsewhere in the world and compared to Dover has many more of the features of the standard waterfront tourism product. Dover has promoted a rather different tourist gaze based on a reconstructed history as a front-line town linked to a wider British nationalism. In contrast, in London Docklands the historical and cultural image of the area is more fragmented and lacks clearly dominant heritage tourist objects alongside attractions, such as Canary Wharf, that are a product of the area's redevelopment.

Despite these differences in approach it is possible to identify a marked similarity in the tourism policies developed in cityport regions. The concern over the tourism–environment relationship mentioned earlier (Shaw and Williams, 1992) does not seem to have been a major influence on tourism policy in the three areas. The recently published five-year objectives for tourism in Dover (Dover District Council, 1994) mentions products, accommodation, marketing and training. In each case study area there are significant policy measures to improve the tourist environment, but these do not seem to be strategically integrated with the economic and social aims of tourism policy.

ACKNOWLEDGEMENTS

My thanks go to colleagues in the Department of Geography, Birkbeck College: Paul Bull for his comments on an earlier draft of this chapter, Tina Scally for the maps, and John Wilson (Photographic Unit) for help with the photographs. Also I am very grateful to staff at Dover Borough Council, Hartlepool Borough Council and the London Docklands Development Corporation for their assistance.

REFERENCES

Anderson, B. (1983), *Imagined communities* (London: Verso).
Ashworth, G. and Voogd, H. (1990), *Selling the city* (London: Belhaven).
Beioley, S., Crookston, M. and Tyrer, B. (1988), 'London Docklands: the leisure element', *Leisure Management* 8(2), 30–3.
Bondi, L. (1992), 'Gender symbols and urban landscapes', *Progress in Human Geography* 16, 157–70.
Booth, P. and Boyle, R. (1994), 'See Glasgow, see culture', in Bianchini, F. and Parkinson, M. (eds.), *Cultural policy and urban regeneration: the West European experience* (Manchester: University of Manchester Press), 182–201.
Boyer, M. (1992), 'Cities for sale: merchandising history at South Street Seaport', in Sorkin, M. (ed.), *Variations on a theme park* (New York: Noonday Press), 181–204.
Boyle, M. and Hughes, G. (1991), 'The politics of the representation of the real discourses from the left on Glasgow's role as European City of Culture', *Area* 23, 217–28.
Brownill, S. (1992), *Redeveloping London Docklands: another great planning disaster?*

(London: Pluto Press).

Buckley, P. and Witt, S. (1985), 'Tourism in difficult areas: case studies of Bradford, Bristol, Glasgow and Hamm', *Tourism Management* 6, 205–13.

Buckley, P. and Witt, S. (1989), 'Tourism in difficult areas, II: case studies of Calderdale, Leeds, Manchester and Scunthorpe', *Tourism Management* 10, 138–52.

Bull, P.J. and Church, A. (1994), 'The geography of employment change in the hotel and catering industry of Great Britain in the 1980s: a sub-regional perspective', *Regional Studies* 28, 13–25.

Bull, P.J. and Church, A. (1995), 'The London Tourism Complex', in Law, C.M. (ed.), *Tourism in major metropolitan areas* (London: Mansell).

Campbell, M. (1990), *Local economic policy* (London: Methuen).

Channel Tunnel Joint Consultative Committee (1987), *Kent impact study* (Maidstone: Kent County Council).

Census of Employment (1981, 1989, 1991), National On-line Manpower Information System, University of Durham, UK.

Church, A. and Reid, P. (1994), 'Anglo-French co-operation: the effect of the Channel Tunnel', in Gibb, R. (ed.), *The Channel Tunnel: a geographical perspective* (London: Wiley), 199–213.

Cleveland County Council (1989), *Visitor survey: results* (Cleveland County Council: Department of Environment, Development and Transportation).

Cleveland County Council (1993a), *Unemployment in Cleveland* (Cleveland County Council: Department of Environment, Development and Transportation).

Cleveland County Council (1993b), *County tourism strategy action programme* (Cleveland County Council: Department of Environment, Development and Transportation).

Cleveland County Council (1993c), *Business tourism: a profile of business visitors to Cleveland* (Cleveland County Council: Research and Intelligence Unit).

Cleveland County Council (1993d), *Urban regeneration initiatives in Cleveland* (Cleveland County Council: Department of Environment, Development and Transportation).

Cleveland County Council (1994), *Visitor and User Survey: Hartlepool Marina*, (Cleveland County Council: Research and Intelligence Unit).

Cooper, C. (1987), 'The changing administration of tourism in Britain', *Area* 19(3), 249–53.

Coopers and Lybrand (1987), *Tourism potential: Teesside* (London: Department of the Environment).

Crilley, D. (1993), 'Megastructures, and Urban Change: Aesthetics, Ideology and Design' in Knox, P.L. (ed.), *The Restless Urban Landscape*, Chapter 6, (New Jersey: Prentice Hall), 127–164.

Department of the Environment (1993), *Planning policy guidance Note 21: Tourism* (London: HMSO).

Dover District Council (1986), *Report to the Planning Committee* (Dover: Dover District Council).

Dover District Council (1987), *Tourism and operational marketing strategy* (Dover: Dover District Council).

Dover District Council (1993), *Report to marketing committee* (Dover: Dover District Council).

Dover District Council (1994), *Tourism business plan 1994/5* (Dover: Dover District Council).

Dover District Council/Eurotunnel (1987), *Dover economic strategy* (Dover: Dover District Council).

Dover Harbour Board, *Annual report* (Dover: Dover Harbour Board, annually).

Dowling, R. (1992), 'Tourism and environmental integration: the journey from idealism to realism', in Cooper, C. and Lockwood, A. (eds), *Progress in tourism, recreation and hospitality management, Volume 4* (London: Belhaven), 33–46.

East Kent Tourism Development Action Programme (1990), *Position statement* (London: Leisureworks).

English Tourist Board (1991), *Tourism and the environment: maintaining the balance* (London: ETB).

Eurotunnel (1986), *Prospects for tourism development initiatives in Thanet, Dover and Shepway* (London: Eurotunnel).

Getz, D. (1989), 'Special events: defining the product'. *Tourism Management* 10, 125–37.

Gilbert, D.C. (1990), 'Conceptual issues in the meaning of tourism', in Cooper, C. and Lockwood, A. (eds), *Progress in tourism, recreation and hospitality management, Volume 2* (London: Belhaven), 4–27.

GLC (1979), *1988 Olympics: feasibility study* (London: Greater London Council).

Hartlepool Borough Council (1993), *Hartlepool Borough Council local plan.*

Hartlepool Borough Council (1994), *A tourism strategy for Hartlepool* (Department of Economic Development and Leisure).

Harvey, D. (1988), 'Voodoo cities', *New Statesman and Society*, 30 September, 33–5.

Harvey, D. (1989), *The condition of postmodernity* (Oxford: Blackwell).

Heeley, J. (1989), 'Role of national tourist organizations in the United Kingdom', in Witt, S.F. and Moutinho, L. (eds), *Tourism marketing and management handbook* (Hemel Hempstead: Prentice-Hall).

Henry, I. (1994), *The politics of leisure policy* (Basingstoke and London: Macmillan).

Hewison, R. (1987), *The heritage industry* (London: Methuen).

Hoyle, B.S. and Pinder, D.A. (1992), 'Urban waterfront management: historical patterns and prospects', in Fabbri, P. (ed.), *Ocean management in global change* (London: Elsevier Applied Science), 482–501.

Hudson, D. (1989), *Wrecking a region* (London: Methuen).

Imrie, R. and Thomas, H. (1993), *British urban policy and the urban development corporations* (London: Paul Chapman Publishing).

Insight Social Research (1991), *LDDC employment census 1990 survey report* (London: Docklands Development Corporation).

Kent County Council (1993), *Population predictions 1993* (Maidstone: Kent County Council).

Kent Ferry Operators (1990), *Annual Report, Kent Ferry Operators* (Dover).

Law, C.M. (1988), 'Urban revitalisation, public policy and the redevelopment of redundant port zones: lessons from Baltimore and Manchester', in Hoyle, B.S., Pinder, D.A. and Husain, M.S. (eds), *Revitalising the waterfront: international dimensions of dockland redevelopment* (London: Belhaven), 20–37.

Law, C.M. (1992), 'Urban tourism and its contribution to economic regeneration', *Urban Studies* 29, 599–618.

Law, C.M. (1993), *Urban tourism* (London: Mansell).

LDDC (1983), *Corporate plan* (London Docklands Development Corporation).

LDDC (1984), *Corporate plan* (London Docklands Development Corporation).

LDDC (1989), *Tourism marketing strategy* (London Docklands Development Corporation).

LDDC (1993), *London Docklands: don't go home without seeing it* (London Docklands Development Corporation).

LDDC (1994), *Visitor strategy update: report to Corporation Board 20th December 1994* (London Docklands Development Corporation).

LDDC (1995), *Toureast London* (London Docklands Development Corporation).

Ledgerwood, G. (1987), *Urban innovation* (Aldershot: Gower).

Lim, H. (1993), 'Cultural strategies for revitalising the city: a review and evaluation', *Regional Studies* 27, 589–95.

Llewelyn-Davies Planning Leisureworks (1987), *Tourism development in Docklands: key themes and facts* (London Docklands Development Corporation).

London Docklands Study Team (1973), *Docklands redevelopment proposals for East London: a report to the Greater London Council and the Department of the Environment. Volume One, Main Report* (London: Department of the Environment).

London Tourist Board (1993), *Tourism strategy for London action plan 1994–7* (London: London Tourist Board).

LPAC (1994), *Planning advice for London* (London: London Planning Advisory Committee).

May, V. and Heeps, C. (1994), 'Coastal zone management and tourism in Europe', in Cooper, C. (ed.), *Progress in tourism, recreation and hospitality management, Volume 1* (London: Belhaven), 132–46.

PACEC (1991), *Kent impact study 1991 review* (Cambridge: PA Cambridge Economic Consultants).

Paddison, R. (1993), 'City marketing', *Urban Studies* 30, 339–50.

Page, S. (1988), 'Tourists arrive in the docks', *Town and Country Planning*, June, 178–9.

Page, S. and Sinclair, M.T. (1989), 'Tourism and accommodation in London: alternative policies and the Docklands experience', *Built Environment* 15(2), 125–37.

Pearce, D. (1988), 'Tourism and regional development in the European Community', *Tourism Management* 9(1), 13–22.

Pearce, D. (1992), *Tourist organizations* (Harlow: Longman).

Peat Marwick McLintock (1988), *The Dover tourism initiative* (Dover: Dover District Council, Peat Marwick McLintock and Research Design Ltd).

Piotrowski, S. and Page, S. (1991), 'Tourism and regeneration – the Dover experience', *Town and Country Planning*, January, 24–26.

Poon, A. (1989), 'Competitive strategies for a new tourism' in Cooper, C. (ed.), *Progress in tourism, recreation and hospitality management, Volume I* (London: Belhaven), 91–102.

Prentice, R. (1994), 'Heritage: a key sector of the "new" tourism' in Cooper, C. (ed.), *Progress in tourism, recreation and hospitality management, Volume I* (London: Belhaven), 309–24.

Robinson, R. and Shaw, K. (1991), 'Urban regeneration and community development', *Local Economy* 6(1), 61–72.

Robinson, R., Shaw, K. and Lawrence, M. (1994), 'Urban Development Corporations and the creation of employment: an evaluation of Tyne and Wear and Teesside Development Corporations', *Land Economy* 8 326–37.

Shaw, G. and Williams, A. (1992), 'Tourism development and the environment: the eternal triangle', in Cooper, C. and Lockwood, A. (eds), *Progress in tourism, recreation and hospitality management, Volume 4* (London: Belhaven), 47–59.

Shaw, G., Greenwood, J. and Williams, A.M. (1988), 'The United Kingdom: market responses and public policy', in Williams, A.M. and Shaw, G. (eds), *Tourism and economic development: western European experiences* (London: Belhaven), 162–179.

Shaw, G. and Williams, A.M. (1994), *Critical issues in tourism: a geographical perspective* (Oxford: Blackwell).

Short, J. (1989), 'Yuppies, yuffies and the new urban order', *Transactions, Institute of British Geographers New Series* 14(2), 173–88.

Sinclair, T. and Page, S. (1993), 'The Euroregion: a new framework for tourism and regional development', *Regional Studies* 27, 475–83.

Sorkin, M. (1992), *Variations on a theme park* (New York: Noonday Press).

TDC (1993), *Annual Report* (Teesside Development Corporation).

Tees and Hartlepool Port Authority (1993), *Annual report*.

Thomas, B. (1995), 'Tourism: is it underdeveloped?' in Evans, L., Johnson, P. and

Thomas, B. (eds), *The Northern Region economy: progress and prospects in the North of England* (London: Mansell), 59–78.

Thomas, C. (1989), 'The roles of historic sites and reasons for visiting', in Herbert, D., Prentice, R. and Thomas, C. (eds), *Heritage sites: strategies for marketing and development* (Aldershot: Avebury), 62–93.

Townsend, A. (1992), 'New directions in the growth of tourism employment? propositions of the 1980s', *Environment and Planning A* **24**, 821–32.

Urry, J. (1990), *The tourist gaze* (London: Sage).

Urry, J. (1993), 'Europe, tourism and the nation state', in Cooper, C. and Lockwood, A. (eds), *Progress in tourism, recreation and hospitality management, Volume 5* (London: Belhaven), 89–98.

Williams, A.M. and Shaw, G. (eds) (1991), *Tourism and economic development: western European experiences* (London: Belhaven Press).

Part IV

INTERMODALISM, MIDAs AND MULTIMODALISM

12 Fixed Links and Short Sea Crossings

RICHARD KNOWLES

Department of Geography, University of Salford, UK

Short sea channels and tidal rivers impose not only a physical barrier to movement but also a transport cost barrier which is out of all proportion to their size (Knowles, 1981). This is due to the high capital costs and terminal costs of waterborne transport (Bird, 1971). Although the real cost of short sea transport fell with the introduction of roll-on/roll-off (ro–ro) vehicle ferry services and has fallen further with the more recent development of multi-deck jumbo ferries, short sea channels and river estuaries remain a substantial transport cost barrier. Gibb (1988), for example, estimated that crossing the 34-km-wide English Channel cost the equivalent of 146 km of land travel in 1988, a 4.3-fold transport cost supplement. This cost barrier depresses the amount of interaction and constrains market areas.

CONSTRAINTS ON FIXED LINK CONSTRUCTION

With specialisation of production in market economies and rising living standards the demand for freight and passenger transport rises. Short sea crossings can then become bottlenecks within the transport system because of limited capacity, intermittent availability and unreliability. The construction of fixed links to eliminate such bottlenecks and reduce the transport cost barrier depends on technical, strategic, demand and safety constraints being surmounted.

Technical constraints

Fixed links can take the form of bridges, tunnels, causeways or pontoons. Although there is no technical limit to the length of viaducts or causeways across shallow water, the current maximum length of an individual bridge span is still under 2 km which is a major constraint for deep-water crossings.

A further constraint is the requirement under international law of free passage in international straits. Navigational clearance is required under bridges across

Cityports, Coastal Zones and Regional Change. Edited by Brian Hoyle.
© 1996 John Wiley & Sons Ltd.

Table 12.1.   Navigational clearance of bridges across international straits

Bridge	Strait/sea	City/country	Navigational clearance (in metres)
Akashi–Kaikyo*	Inland Sea	Honshu–Shikoku, Japan	65
Bosporus	Bosporus, Black Sea	Istanbul, Turkey	62
Golden Gate	Golden Gate, Pacific Ocean	San Francisco, USA	64
Great Belt East*	Great Belt, Baltic Sea	Korsør, Denmark	65
Lions Gate Bridge	First Narrows, Georgia Strait	Vancouver, Canada	60
Øresund Bridge**	Øresund, Baltic Sea	Copenhagen, Denmark, Malmö, Sweden	57

* = under construction.
** = at pre-tender stage.

Sources: Stærbo (1989), Øresund konsortiet (1994).

international straits to allow passage of the largest ships in existence at the time. Furthermore, such bridges must be constructed so as not to impede the passage of ships (Stærbo, 1989). Most bridges over international straits have navigational clearances above 60 m (Table 12.1). Such elevated structures are more expensive to build and often also to use as vehicles will use more fuel due to the steepness of the approach road gradient of the bridge unless it is built from land well above sea level. The Great Belt East Bridge in Denmark, for example, will have a relatively steep gradient of about 25 per 1000 whereas the maximum permitted gradient on Danish motorways is 35 per 1000. A bridge across the Messina Strait between Italy and Sicily, with a free span of 3300 m, is now approaching the technical limits of feasibility (Holmegaard, 1995). A United Nations Working Group has evaluated bridge projects for the 14-km-wide Strait of Gibraltar at the mouth of the Mediterranean Sea between Europe and Africa, although a tunnel is more feasible technically.

The three principal types of bridge are arch bridges, girder or beam bridges, and suspension and cable-stayed bridges (Figure 12.1). By far the greatest spans are achieved by suspension bridges with the 1410-m-span Humber Bridge in England, opened in 1981, the longest (Table 12.2). This will soon be exceeded by the 1624-m-span Great Belt East Bridge in Denmark in 1998 and then the 1990 m-central-span Akashi–Kaikyo Bridge between the islands of Honshu and Shikoku in Japan later in 1998. The greatest single span arch bridge is the 518-m New River Gorge Bridge in West Virginia, USA, opened in 1977, while the maximum length of girder bridges is much shorter. The longest pontoon or floating bridge is

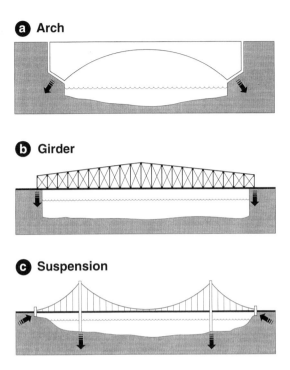

Figure 12.1. The three principal types of bridge

the 2291-m floating section of the Second Lake Washington Bridge in Seattle, USA, opened in 1963. Although the technical limits are less severe on building long tunnels through mountains or under sea channels or rivers, or on immersing pre-fabricated tunnels on the sea bed, the longest tunnels are the 53.9-km Seikan Tunnel between the islands of Honshu and Hokkaido in Japan which was opened in 1988 and the 49.9-km Channel Tunnel between England and France, opened in 1994, of which 31 km is under the sea.

Ventilation is the major technical constraint on tunnel length, especially for under-sea tunnels over 4 km in length which can only be currently utilised by electrically driven trains. The western section of the 16-km Danish–Swedish Øresund Fixed Link, a 3.75-km submerged tunnel containing a two-lane motorway and double-track high-speed rail line, due to open in the year 2000, reflects the current limits for under-sea road tunnels (Sund and Bro, 1994). Exhaust gases from both petrol and diesel engine vehicles present an insurmountable problem for the ventilation of long under-sea tunnels. Tunnels are therefore a less attractive form of fixed link than bridges as road transport requires expensive ventilation systems in short tunnels and can only use long under-sea tunnels by piggybacking on railway trains. Only the development of electric cars will overcome this major ventilation problem.

Table 12.2. Longest fixed links by category

Category	Location, country	Single span length (metres)	Year of opening
Steel arch bridge	New River Gorge, West Virginia, USA	518	1977
Cantilever bridge	Pont de Québec, St Lawrence River, Canada	549	1917
Girder (beam) bridge	Numerous, e.g. Great Belt West Bridge, Denmark	110	1995[1]
Cable suspension bridge	Humber Bridge, England	1410	1981
Cable stayed bridge	Pont de Normandie, Le Havre, France	856	1994
Floating bridge	Second Lake Washington Bridge, Seattle, USA	2291	1963
Causeway	Second Lake Pontchartran Causeway, Louisiana, USA	38 422	1969
Viaduct	Great Salt Lake Railroad Trestle, Utah, USA	19 000[2]	1904 (1960)[3]
Under-sea rail tunnel	Channel Tunnel, England–France	31 000[4] (49 940)[2]	1994
	Seikan Tunnel, Honshu–Hokkaido, Japan	23 300[4] (53 850)[2]	1988
Mountain rail tunnel	Simplon II, Switzerland–Italy	19 820	1922
Urban rail tunnel	Kaluzhskaya Metro, Moscow, Russia	37 900	1990
Mountain road tunnel	St Gotthard, Switzerland	16 320	1980

[1] Not in use until Rail Tunnel and/or East Bridge open 1997/98.
[2] Total length.
[3] Converted to rock fill.
[4] Under-sea section.

Sources of data: Marsden (1993), Matthews (1994).

Strategic constraints

Fixed links have sometimes not been developed primarily for strategic reasons. The Channel Tunnel between England and France, for instance, was repeatedly vetoed by the British Government in the 19th and early and mid-20th centuries for fear of providing an easy invasion route for foreign armies from mainland Europe (Gibb, 1994). These strategic objections were only abandoned in 1955 after missile and airborne warfare technology reduced the military significance of this international fixed link. However, where fixed links replace rather than complement short ferry routes, there is a loss of flexibility as ferries can be diverted to operate from different ports after a port has been closed due to an accident or military strike. In contrast, fixed links are more vulnerable and are very difficult to replace quickly. It was for similar strategic reasons that the US Interstate Highway System, first proposed as the National System of Interstate and Defense Highways in 1944, was designed as a transport network with high connectivity ratios and a large number of circuits or alternative routes so that the elimination of individual nodes would not paralyse the movement of military supplies (Taaffe and Gauthier, 1973).

Transport demand

Most fixed links are only built when the growth in transport demand is sufficient to pay back through user tolls a loan for construction at commercial rates of interest. User tolls for fixed links are often set at the same level as the previous short sea ferry service on the premise that consumers are gaining sufficient benefit from the fixed link's higher capacity, 24-hour availability and faster speed of traffic. This can result in high and widely varying tolls per kilometre. The Skye Bridge linking the Isle of Skye to the Scottish mainland from 1995 has a £5.20 single charge for cars for a 1-km crossing, the highest such charge in the European Union, whereas the 3-km Forth Road Bridge near Edinburgh (Figure 12.2) charges only 40p per car (*The Guardian*, 15 July 1995, p. 6). Where a fixed link competes with rather than replaces a ferry service, price cutting can benefit the consumer at

Figure 12.2. Suspension road bridge (left) and cantilever rail bridge (right) over the Forth Estuary, Scotland. (R.D. Knowles)

least in the short term as the rival modes seek a larger market share. Such competition is already emerging between the Channel Tunnel and short sea English Channel ferries. However, while any user tolls are charged the short sea cost barrier remains.

Some fixed links which are not commercially viable are built for prestige, political or economic development reasons. For example, the Humber Bridge near the city of Hull in England was built as a result of a Government commitment made during a parliamentary by-election in Hull North constituency in 1966. The bridge, which is not part of an inter-regional route and serves only a local traffic function, carries insufficient traffic for the toll income to even pay off the interest on the construction loan so that the total debt is rising annually. In the remote and depopulating Outer Hebrides islands of Scotland however, the Government paid for toll-free causeways to link North Uist to Grimsay and Benbecula, a toll-free bridge to link Benbecula to South Uist and modern subsidised ro–ro vehicle ferry services linking the islands with each other and the Scottish mainland (Knowles, 1981). These transport developments, which lower the island transport cost barrier and help stimulate local economic development, have been introduced for a mixture of social, party political and strategic reasons.

Safety constraints

The expense of fixed links is increased by the necessary requirement for high design safety and operational standards. Procedures for evacuation in the case of fire, explosion, flooding or operational accidents are particularly onerous for long under-sea tunnels. In the case of the Channel Tunnel a third service tunnel provides both a maintenance and evacuation route for the two railway tunnels. Bridge design has to minimise the danger of collision from shipping and aircraft. With increasingly stringent safety standards and the increased risk of terrorist attack, safety constraints are becoming an increased obstacle to fixed link construction.

EVOLUTION OF SHORT SEA LINKS

Short sea shipping has rapidly evolved. Fifty years ago, conventional liners linked often quite distant traditional ports with low frequency services. These ports were major employers due to the highly labour-intensive methods of loading and unloading freight and vehicles, and the processing of raw materials and manufacturing provided considerable employment in port-side industries (Bird, 1971; Knowles, 1981; Willingale, 1981; Gibb, 1987). Ports were terminals and generated urban growth (Table 12.3; Figure 12.3a). With the introduction of roll-on/roll-off multipurpose vehicle ferries, port employment and handling costs were reduced and more direct port alignments were sought with shorter sea crossings (Table 12.3; Figure 12.3b). Ports were changing from being terminals

Table 12.3. Evolution of short sea links

Short sea link	Port link employment	Technology
A Ports – terminal concept	High	Conventional ships/ traditional ports
B Ports – through concept	Low	Ro–Ro ferries and container ships
C Fixed link – tunnel or bridge	Minimal	Drilling Ventilation Bridge Structure

and major sources of employment in a sea transport system to becoming merely links in a multimodal transport system connecting regions and countries (Figure 12.4).

The development of larger, faster roll-on/roll-off ferries, often multi-deck with simultaneous loading, encouraged the development of shortest sea crossings which are not always associated with traditional ports. This maximised the length of faster land transport and minimised the relatively slower sea transport within the overall transport system. This reduced through journey times and made more intensive use of the large amount of capital invested in ferries (Knowles, 1981, 1994). The construction of fixed links, whether tunnels or bridges, as replacements for short sea crossings (Table 12.3; Figures 12.3c and 12.5) is merely a further development of the through transport concept made possible by advances in tunnel drilling and ventilation technology and the technology of bridge structures and materials.

Fixed links reduce port employment considerably while providing only minimal permanent employment in maintenance and toll collection after the substantial temporary employment during construction. However, fixed links reduce transport costs, enhance transport capacity and allow market areas to be widened generating an overall net benefit to the economy. An example of the employment impacts of fixed links on ports is the estimated effects of the Channel Tunnel on East Kent's ports. In the mid-1980s, it was predicted that depending on port traffic levels, East Kent port employment would fall from 12 450 to between 6600 (−47%) and 9500 (−24%) in the first year of Channel Tunnel operation (Table 12.4) (Channel Tunnel Joint Consultative Committee, 1986). With some recovery in ferry traffic volumes by 2003, port employment would recover to between 7400 (−41%) and 11 300 (−9%). Estimated Channel Tunnel employment of 3200 in its first year of operation would leave an overall deficit of between 1100 and 3400 jobs. By 2003, an increase in Channel Tunnel employment to 4500 and partial recovery in port employment to between 7400 and 11 300 would leave overall employment levels between 2100 lower and 400 higher than in the mid-1980s, but handling twice as much traffic.

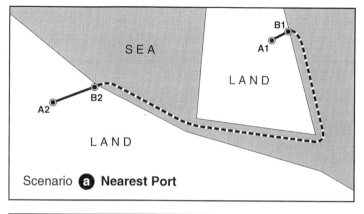

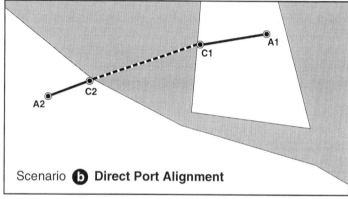

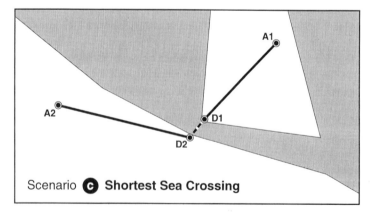

Figure 12.3. The evolution of short sea links

Figure 12.4. Øresund ro–ro ferry at Helsingør ferry port, Denmark. (R.D. Knowles)

Figure 12.5. Suspension road bridge over a deep-water fjord near Bergen, Norway which replaced a ro–ro ferry route. (R.D. Knowles)

Table 12.4. East Kent port employment

	1985	1993*	2003
No tunnel	12450	13200–13800	14000–15400
With tunnel			
(a) Port	12450	6600–9500	7400–11300
(b) Tunnel	0	3200	4500
Net change	0	−3400/−1100	−2100/+400

* Assumed to be Year 1 of Channel Tunnel operation.

Source: Channel Tunnel Joint Consultative Committee (1986).

LAND ROUTE CONCENTRATION

The replacement of numerous low frequency shipping routes between conventional ports by cheaper and faster high frequency shortest sea ferry routes, and then sometimes in turn by fixed links, encourages a concentration of land routes on the shortest sea port or fixed link. These ports and fixed links are new transport hubs and become nodal points in transport systems which can be of international, national, regional or just local scale. They acquire the spatial quality of 'intermediacy', 'gateway' locations between important sets of origins and destinations, which become more significant with the increased use of more efficient intermodal transport (Bird, 1983; Fleming and Hayuth, 1994).

TRAFFIC DIVERSION AND GENERATION

Fixed links attract traffic away from other modes of transport, principally from ferries when they complement or replace short sea services. Fixed links also generate additional traffic. Existing short sea passenger traffic is mainly leisure traffic which is not as sensitive as business traffic to journey time. Fixed links enable cars, coaches and/or trains to operate at any time of the day or night, irrespective of the weather. Traffic capacity, frequency, reliability and journey speed are all enhanced by fixed links. Competing ferries can respond by lowering prices, by repackaging the ferry journey as a short sea cruise or by marketing the ferry trip as a driver's rest break. This is particularly important for coach and lorry drivers who in the European Union have mandatory drivers' hours rules including specified rest breaks. For international journeys, short sea ferries have additional attractions not available to international train services of on-board duty-free sales of highly taxed products such as alcohol, tobacco and perfume. Within the European Union however, this is a short-term advantage which should have been abolished with the creation of the Single Market in 1993 and has been reprieved only until 1999. In the case of cross-English-Channel ferries, duty-free sales bring in more revenue than ferry ticket sales (Knowles, 1994).

After 1999 therefore, international ferry ticket prices within the European Union are likely to increase sharply and where there are competing fixed links more traffic is likely to be attracted away from short sea ferries. Estimates of potential passenger and car traffic diversion away from competing ferries to Eurotunnel's Le Shuttle trains vary between 40 and 60% of cross-Channel ferry traffic to and from Kent ports and between 20 and 37% to and from other English south coast ports, excluding Plymouth which is too distant to be affected (Knowles, 1994).

Århus, Denmark's second largest city, is currently just north of the dividing line across the Jutland peninsula below which the Great Belt rail and car ferry services are more convenient for passengers to and from Copenhagen and the rest of East Denmark than the Kattegat car ferry services. The current modal split is Great Belt 43% (train ferry 32%, car ferry 11%), Kattegat ferries 46% and air transport 11% (Stærbo, 1994). After the Great Belt Fixed Rail and Road Links open in 1997/98 the estimated modal split is Great Belt 72% (train 44%, car 28%), Kattegat ferries 21% and air transport 7%. About 54% of the Kattegat ferry passengers are therefore predicted to divert to the Great Belt Fixed Link.

Where fixed links enable high speed intercity rail transport to compete with air transport, a diversion rate of more than 50% of air traffic to rail is possible. This includes much of the highly priced and highly profitable business traffic but diversion rates depend on speed, overall journey time and price. SNCFs (French Railways) experience with domestic high speed TGV trains is that 92% of air passengers divert to rail if the overall journey time between city centres is 2 hours or below, 70% if 3 hours and 50% if 5 hours (Smalley, 1993). Eurostar intercity trains between London and Paris and London and Brussels via the Channel Tunnel are unlikely to divert more than 50% of the corresponding air traffic as ticket prices are higher and speeds are lower than on French TGV services. Higher ticket prices reflect the higher construction costs of triple-voltage international Eurostar trains operating on the rail systems of three different countries. The lower speeds are mainly the result of delays in opening the Union Railway (London St Pancras to Channel Tunnel High Speed Line) until at least 2002 and also the TGV Belge (Belgian High Speed Line) from Lille to Brussels until 1996/97 (Table 12.5). Then a further diversion of air traffic to high speed rail is likely with the London–Brussels journey time reduced to 2 hours 10 minutes non-stop and the London–Paris journey time reduced to 2 hours 30 minutes non-stop, both faster than air transport's city centre to city centre travel time (Knowles, 1994).

Fixed links generate additional traffic across short sea crossings as they provide faster, more frequent, more reliable and sometimes cheaper journeys than ferries. For example, car traffic increased by 68% and lorry traffic by 14% when the 14-km long Honshu–Shikoku Bridge replaced a short sea ferry service in Japan in 1988. Similarly when the 3-km long Severn Bridge replaced a short sea ferry service between England and Wales in 1966, car traffic increased by 68% and lorry traffic by 14%. When the Great Belt Fixed Link opens in 1997/98 between the islands of Zealand in East Denmark and Fyn in West Denmark, car

Table 12.5. Comparative high speed rail journey times

From	To	Year	Kilometres	Rail time (hours/minutes)	Kilometres per hour
Paris Gare de Lyon	Lyon Part Dieu	1982	422	2.00	211
London Waterloo	Paris Gare du Nord	1994	493	3.00	164
London Waterloo	Brussels Midi	1994	374	3.10	118
London Waterloo	Brussels Midi	1996/97 *	c.374	2.40	140
London St Pancras	Paris Gare du Nord	2002/03 **	c.493	2.30	197
London St Pancras	Brussels Midi	2002/03 **	c.374	2.10	173

* = Opening of TGV Belge.
** = Opening of Union Railway.
Sources of data: Knowles (1994); Smalley (1993).

traffic is predicted to be 30% higher than on the ferry service which it directly replaces. This is a rather lower level of traffic generation because competing ferry services will continue to operate after 1997/98 on other routes between East and West Denmark.

HINTERLAND EXPANSION AND COMPETITION

Fixed links reduce the friction of distance and increase what Ullman (1956) defined as 'transferability' by increasing traffic speed and capacity and often by lowering both transport costs and more significantly charges to transport users. As a result, short sea straits and tidal rivers become a less significant barrier to the movement of people, raw materials and finished products. Market areas can widen more easily beyond a fixed link than beyond a short sea ferry crossing. Economic specialisation, the volume of interaction, trade patterns, hinterlands, urban hierarchies and even commuting fields can all be affected by the opening of a fixed link.

Murayama (1994) demonstrated the impact of railways on accessibility on the Japanese urban system. The urban system, traditionally linked by coastal shipping and a primitive road system, 'was completely reorganised by the emergence of railways which created new connections and greatly increased the speed of existing links' (Murayama, 1994, 87). Fixed links to the west from the main island of Honshu – to the islands of Kyushu by rail tunnel and later by bridge, to Shikoku by bridge and to the north to the island of Hokkaido by rail tunnel – eliminated the need for ferries (Figure 12.6). Economic integration was further enhanced by the development of high speed Shinkansen train routes. Travel time in Japan has reduced by 80% since 1900. The Shinkansen cities have gained most in locational advantage, especially the capital Tokyo which is the terminus of all three Shinkansen routes. Tokyo's metropolitan hinterland has widened dramatically.

The Øresund road and rail link (see Figure 12.7) between Copenhagen, on the eastern Danish island of Zealand, and Malmö in southern Sweden, which is due to open in the year 2000, offers new growth potential to Copenhagen according to Matthiessen (1993). The Øresund Strait is a time and price barrier with different languages, cultures and economies on either side. Swedish membership of the European Union from 1995 and the Øresund Fixed Link from 2000 will enhance the growth potential of the Copenhagen (1.7 million population) and Malmö–Lund (0.5 million population) conurbations. Matthiessen (1993) argued that the new Danish–Swedish agglomeration could become the centre of a new South Scandinavian region of 8 to 9 million inhabitants resulting in large-scale changes in the dominance and hinterland patterns of the respective national capitals Copenhagen and Stockholm. New synergy in the Øresund conurbations could revive the 1960s concept of Øresundstad, a single Øresund urban agglomeration (Fullerton and Williams, 1972). Matthiessen demonstrated how the dramatic accessibility changes of the Øresund and Great Belt Fixed Links should extend the catchment area and function of Copenhagen Airport which is already the main European international hub north of Frankfurt and Amsterdam. The Great Belt Fixed Link will from 1997 bring Copenhagen nearer by rail than air in travel time to all provincial Danish cities except for Ålborg in northern Jutland. Faster and cheaper journeys by rail should result in a significant mode transfer away from air transport for domestic traffic as Copenhagen casts its traffic shadow over most of Denmark. The Øresund Fixed Link should redirect to Copenhagen by rail or road much of the south Swedish international air traffic which currently connects to Stockholm. National links and greater distances between urban areas in Sweden will enable links between south Swedish airports and Stockholm to survive at a lower level. The Great Belt and Øresund Fixed Links should therefore respectively extend Copenhagen's influence area in West Denmark in competition with the city of Odense on the island of Fyn and into southern Sweden in competition with the cities of Malmö and Lund. This could affect the urban hierarchy in southern Scandinavia, stimulating both Copenhagen's urban growth as the principal urban centre and some international commuting.

A third Scandinavian Fixed Link, across the Fehmarn Belt between the East Danish island of Lolland and the German island of Fehmarn, is under investigation. This combined road and rail bridge and/or tunnel project with associated motorway and high speed rail development would provide a much faster link from East Denmark, Sweden and Norway to Germany and the core of the European Union. The spatial development impacts of the Fehmarn Belt Fixed

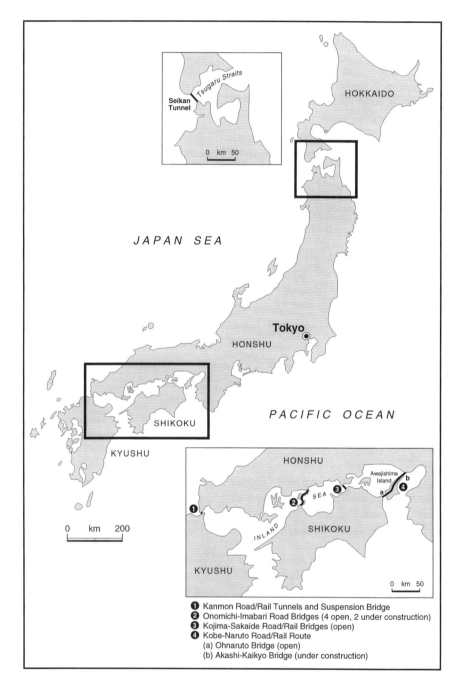

Figure 12.6.   Japan: inter-island fixed links

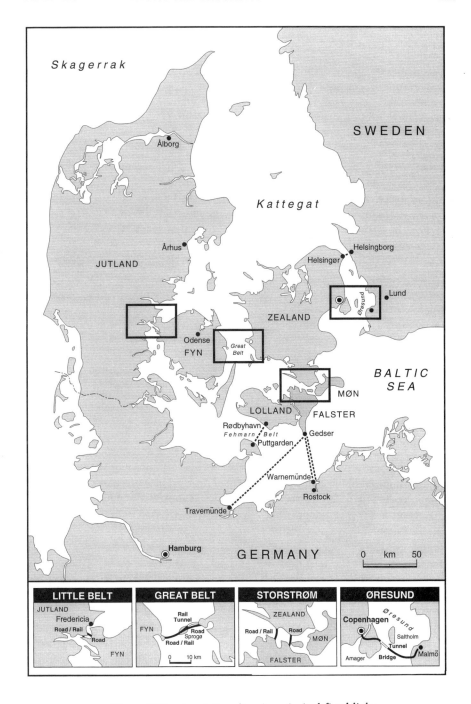

Figure 12.7. South Scandinavia: principal fixed links

Link are complex as it would extend and overlap the influence areas of Hamburg and Copenhagen while diverting most East Danish, other Scandinavian and German international traffic from the more circuitous Great Belt Fixed Link route (Storebælt, 1990; Danish Transport Council, 1995). With new high speed trains, rail travel would be possible from Stockholm to Hamburg via the Fehmarn Belt Fixed Link in 5 hours compared with 16 hours today and 10–12 hours via the Øresund and Great Belt Fixed Link from 1997 (Sund and Bro, 1995). The Fehmarn Belt Fixed Link would replace the Rødbyhavn–Puttgarden ferry route and absorb its traffic, while competing with the other three Danish–German ferry routes from Gedser, on the Danish island of Falster, to Warnemünde, Travemünde and Rostock.

Views differ widely on the possible effects of the Channel Tunnel on regional development and urban systems and what Button (1994) defines as the 'directionality' of benefits (Figures 12.8, 12.9, 12.10 and 12.11). The short-term stimulus of engineering orders and employment during the construction phase appears to have been concentrated in the neighbouring Nord/Pas de Calais region in France but more dispersed throughout Britain. France sees the Channel Tunnel providing longer-term regional economic development benefits spreading through northern France and especially the highly accessible Lille region with its high speed rail megahub (Button, 1994; Knowles, 1994). A hundred TGV trains will stop daily at Lille alongside a major new office, conference and retail complex, including 22 Eurostar and 10 Eurostar Beyond London trains carrying a total of 30 million passengers each year (Perren, 1992; Webster, 1993).

In Britain, accessibility gains are concentrated on London and the South East. Wider regional implications of the Channel Tunnel are likely to be restricted by the limited infrastructural investment programme in regional freight terminals, the absence of a High Speed Line from London to the Tunnel until at least 2002, and the handful of regional through passenger services (Gibb et al., 1992). Traffic flows are, however, already buoyant on the daily return intermodal freight trains from the new Manchester Trafford Park Euroterminal to Paris and Milan. During a recent field visit to the Terminal, the Manager Mr D.K. Bass revealed that unexpected traffic includes the export of pizza bases to Italy and English bread to Paris!

Improved transport links can also assist the development of rural areas. In the Outer Hebrides, off the west coast of Scotland, fixed links and inter-island ro–ro vehicle ferries enabled these islands, for the first time in 1975, to form one single-tier government unit based in Stornoway with a district office in Benbecula (Knowles, 1981). Previously the islands were administered from either Dingwall or Inverness on the east coast of Scotland. Also all secondary school pupils could now be educated on the islands for the first time. The transferred employment helped to boost Stornoway's near-static population from 5152 in 1971 to 5656 in 1979 and Benbecula's from 1355 to 1789, and stimulated demand for inter-island transport.

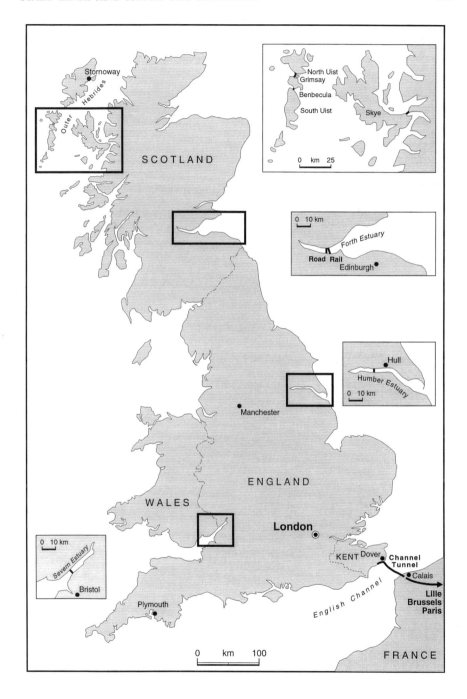

Figure 12.8. Great Britain: selected fixed links

Figure 12.9.   Channel Tunnel entrance at Coquelles, France. (R.D. Knowles)

Figure 12.10. Cutting head from a Channel Tunnel boring machine, outside the Eurotunnel Exhibition Centre, Coquelles, France. Its size shows the huge dimensions of both single-track operating tunnels which allow a much larger loading gauge and carrying capacity for Le Shuttle trains than on conventional railway systems. (R.D. Knowles)

Figure 12.11. Channel Tunnel rock waste tip forming a new walled platform at the foot of Shakespeare Cliff, Kent. This contains the rock waste from the British end of the two operating tunnels and the service tunnel. (R.D. Knowles)

CONCLUSION

Fixed links often replace short sea ferries where technical, strategic and safety constraints can be overcome. As the development of fixed links is usually demand-led, they are more prevalent on principal inter-urban, international and intra-urban routes. They represent a later stage than ro–ro ferries in the development of the through transport concept. Land routes converge on fixed links which divert traffic away from other modes. Fixed links also generate additional traffic because of their round-the-clock availability, additional capacity and higher traffic speeds. Market areas can widen more easily beyond a fixed link than beyond a short sea ferry crossing. Fixed links are therefore potent forces in helping to reshape spatial development in port-city regions and can enhance the traffic and economic shadow cast by metropolitan areas.

REFERENCES

Bird, J.H. (1971), *Seaports and seaport terminals* (London: Hutchinson).

Bird, J.H. (1983), 'Gateways: slow recognition but irresistible rise', *Tijdschrift voor Economische en Sociale Geografie* 74(3), 196–202.

Button, K. (1994), 'The Channel Tunnel and the economy of southeast England', *Applied Geography* 14, 107–21.

Channel Tunnel Joint Consultative Committee (1986), *Kent impact study: a preliminary assessment* (Maidstone: Kent County Council).

Danish Transport Council (1995), *Facts about Fehmarn Belt* (Copenhagen: Danish Transport Council).

Fleming, D.K. and Hayuth, Y. (1994), 'Spatial characteristics of transportation hubs: centrality and intermediacy', *Journal of Transport Geography* 2(1), 3–18.

Fullerton, B. and Williams, A.F. (1972), *Scandinavia* (London: Chatto & Windus).

Gibb, R.A. (1987), 'The major problems and opportunities of a cross-channel rail link', in Tolley, R.S. and Turton, B.J. (eds) *Short sea crossings and the Channel Tunnel* (IBG Transport Geography Study Group), 81–110.

Gibb, R.A. (1988), 'Geographic implications of the Channel Tunnel', *Geography Review* 2(2), 2–7.

Gibb, R.A. (1994), 'The Channel Tunnel project: origins and development' in Gibb, R.A. (ed.), *The Channel Tunnel: a geographical perspective* (London: Wiley), 1–30.

Gibb, R.A., Knowles, R.D. and Farrington, J.H. (1992), 'The Channel Tunnel rail link and regional development: an evaluation of British Rail's procedures and policies', *Geographical Journal* 158(3), 273–85.

Holmegaard, K. (1995), 'Messina and Gibraltar', *News from Storebælt*, No. 3/95, 10–11.

Knowles, R.D. (1981), *Island to mainland transport development in highland areas with reference to the Hebridean Islands of Western Scotland* (Discussion Papers in Geography 12, Department of Geography, University of Salford).

Knowles, R.D. (1994), 'Passenger transport and the Channel Tunnel', in Gibb, R.A. (ed.), *The Channel Tunnel: a geographical perspective* (London: Wiley), 55–77.

Marsden, H. (ed.) (1993). *Whitakers Almanack 1994* (London: J. Whitaker and Sons Ltd).

Matthews, P. (ed.) (1994), *The New Guinness Book of Records 1995* (London: Guinness Publishing Ltd).

Matthiessen, C.W. (1993), 'Scandinavian links: changing the pattern of urban growth and regional air traffic', *Journal of Transport Geography* 1(2), 119–24.

Muruyama, Y. (1994), 'The impact of railways on accessibility in the Japanese urban system', *Journal of Transport Geography* 2(2), 87–100.

Øresund konsortiet (1994), *The fixed link across Øresund* (Copenhagen and Malmö: Øresunds konsortiet).

Perren, B. (ed.) (1992), *Rails into Europe* (Croydon: Eurostar).

Smalley, R. (1993), 'Prospects for passenger services to regions beyond London', presentation on behalf of Union Railways to the Public Transport Information Unit's Seminar, *The Channel Tunnel: ending the North–South divide* (London, 19 July).

Stærbo, C. (1989), 'Navigational clearance of 65 metres for the East Bridge', *News from Storebælt* No. 6/89, 2–4.

Stærbo, C. (1994), 'New model for east–west traffic', *News from Storebælt* No. 01/94, 2–8.

Storebælt (1990), *Fixed links across the Sound and the Fehmarn Belt – Traffic and Economy* (Copenhagen: A/S Storebæltsforbindelsen).

Sund and Bro (1994), Information on the Swedish/Danish Øresund Link, *Sund and Bro* No. 5, June.

Sund and Bro (1995), 'Inauguration of X2000 Stockholm to Malmö train', *Sund and Bro*

No. 7, March, 6–7.

Taaffe, E.J. and Gauthier, H.L. (1973), *Geography of transportation* (Englewood Cliffs, New Jersey: Prentice-Hall).

Ullman, E.L. (1956), 'The role of transportation and the bases for interaction', in Thomas, W.L. (ed.), *Man's role in changing the face of the earth* (Chicago: University of Chicago Press), 862–80.

Webster, P. (1993), 'High-speed French leave BR standing', *The Guardian*, 19 May.

Willingale, M. (1981), 'The port-route behaviour of short sea ship operations: theory and practice', *Maritime Policy and Management* 8(2), 109–20.

13 The Venice Port and Industrial Area in a Context of Regional Change

STEFANO SORIANI

Department of Economics, University of Venice, Italy

The concentration of Italian industrial development in coastal zones, as a result of the combined effects of the country's geography and its heavy dependence upon external sources of raw materials, was well established by the beginning of the 20th century (Zanetto, 1989). Venice exemplifies this pattern, with the expansion of port facilities from the islands towards the mainland and the creation of the industrial zone of Porto Marghera during the early decades of this century. The need to expand port facilities at that time arose from the new vitality shown by the port of Venice as a reflection of the buoyant state of the Italian economy. There was, however, an unfortunate mismatch between the positive and encouraging prospects for port development and traffic growth in Venice in general terms and the structural inadequacy of the relatively new Marittima zone. Opened for traffic in 1880 in the western part of the city, the Marittima zone handled 90% of the total throughput of the port of Venice by the turn of the century (Toniolo, 1972).

The changing balance between the islands and the margins of the Venetian lagoon in terms of port activity was introduced in the early years of this century by the drainage of the Bottenighi sandbank and the excavation of the direct Vittorio Emanuele canal connecting the Marittima zone with the new port on the mainland, where the provision of electricity in 1917 marked the beginning of the process leading to industrial concentration on the margins of the Venetian lagoon. Originated by the 1925 General Plan, the first industrial zone was not completed in 1946 but, extending over 500 hectares, employed 17 000 people in 1940. The chief industries involved were coal distillation, the production of sulphuric acid, pesticides manufacture, shipbuilding, oil refining and glass manufacture. More elaborate plans in 1953 and 1956 introduced petrochemical industries, electricity production and alimentary oil industries in a second industrial zone. With a total area of 1500 hectares, Porto Marghera reached a peak employment level of 33 000 in 1965 and represented one of the most

Cityports, Coastal Zones and Regional Change. Edited by Brian Hoyle.
© 1996 John Wiley & Sons Ltd.

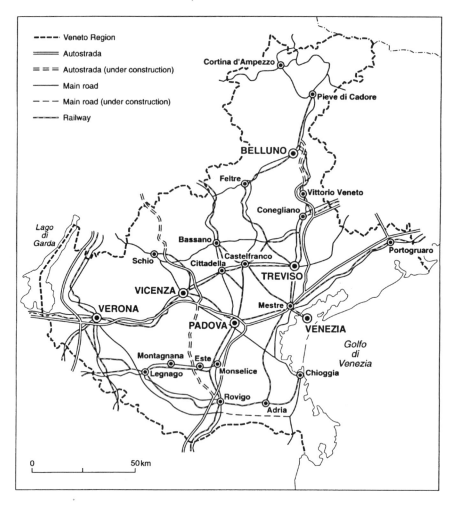

Figure 13.1. The Veneto Region, Italy

important examples of coastal zone industrial development in Italy (Ente Zona Industriale di Porto Marghera, 1990).

PORT–INDUSTRIAL DEVELOPMENT AND THE VENETIAN LAGOON

Port–industrial development in the Venice area, especially during the post-Second World War period, has involved environmental impacts of great significance. Industrial waste has strongly contributed to the modification of the lagoon's ecosystem. Dredging operations, especially those designed in the 1960s to facilitate the access of increasingly large vessels, have seriously affected the morphology of the lagoon and its hydrological regimes. The introduction of coastal barriers to protect the entrances to the lagoon and the excavation towards the end of the 1960s of an 'oil canal' designed to allow large tankers to enter the port without passing the city's historic core, have also increased the influence of marine forces on the dynamics of the lagoon. In addition, the abstraction of ground water for industrial purposes appears to have contributed to the slight but significant gradual reduction in the height of Venice above sea level (Carbognin et al., 1993).

Porto Marghera and the evolution of the Venice area

The expansion of port facilities towards the mainland – the only possible solution if Venice was to develop as a major port–industrial complex – clearly changed the nature of the relationships between the ancient city and its local context. The city in the lagoon, hitherto self-contained, became the historical centre of a Venetian conurbation progressively expanding then becoming more specialised in functional terms (Costa et al., 1971). The evolution of administrative boundaries illustrates this change. In the first three decades of this century the plan for a 'greater Venice' reflected port–industrial growth and was based on the principle that, in administrative terms, Venice should continue to include its port (Chinello, 1979). Marghera, a small rural village required by the industrial initiators to meet the housing demands brought about by rising port–industrial employment, was formally annexed to the Venice Commune in 1917. The same happened to Mestre in 1926. In the enlarged commune resulting from these changes, Venice maintained a central administrative role in comparison with a mainland that was emerging as an industrial suburb.

The changing demographic balance between the historic core and the mainland suburbs underlines the polarising effects of the creation of Porto Marghera. During the period 1920 to 1950 population growth in the mainland part of the commune, which rose from 25 000 to 97 000 was mainly due to in-migration from surrounding rural areas characterised by a structural lack of employment opportunities (Costa et al., 1971; Ente della Zona, 1990). Later on, during the period 1951–69, the mainland population rose from 97 000 to 202 000, primarily as a result of a demographic decrease from 190 000 to 127 000 in the historical centre (Costa et al., 1971). New jobs provided by the growing industrial pole at Porto Marghera and the difficulty of meeting an increasing demand for working class housing in the historic core of Venice led to a marked population decrease in this area, later described as an exodus. Until 1968 population growth in the mainland zone showed an annual increase greater than the decrease recorded in the historical centre. Additional migrants came from a wider hinterland, underlining the economic vitality of the area which offered more employment opportunities than Venice as a whole could itself absorb (Costa, 1987).

The new technological and spatial requirement of the growing port–industrial complex encouraged a process of relocation of industrial activities towards the

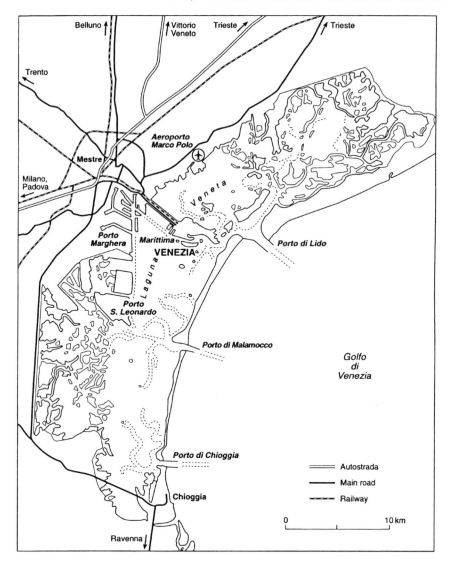

Figure 13.2.   The Venice lagoon and Porto Marghera, Italy

mainland. With the development of Porto Marghera, it was in fact impossible to keep any manufacturing industry in the historical centre of Venice. The migration of the centre of gravity of port activities to the mainland removed the last remaining industrial location advantage from the lagoon. Many traditional industrial activities including the Arsenal, the Stucky mill, the naval dockyards on Giudecca Island and many glassworks on Murano, were abandoned as a result of the impossibility of maintaining on lagoon sites activities dependent on the

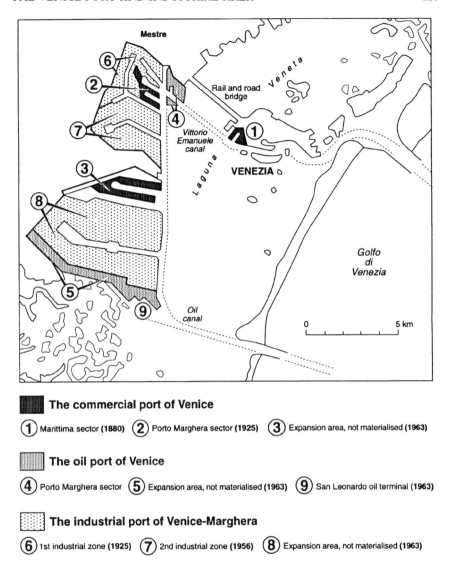

The commercial port of Venice

(1) Marittima sector (1880)   (2) Porto Marghera sector (1925)   (3) Expansion area, not materialised (1963)

The oil port of Venice

(4) Porto Marghera sector   (5) Expansion area, not materialised (1963)   (9) San Leonardo oil terminal (1963)

The industrial port of Venice-Marghera

(6) 1st industrial zone (1925)   (7) 2nd industrial zone (1956)   (8) Expansion area, not materialised (1963)

Figure 13.3.   Port and industrial development in the Venice lagoon, Italy

handling of large quantities of materials. Such activities moved to the mainland where more attractive locations were available, with more ample space and new infrastructures, less competition with other urban functions and the possibility of establishing positive interactions within the developing industrial complex.

Port–industrial development at Porto Marghera clearly required new transport infrastructures aimed at putting to good use the locational advantage offered by the Venice lagoon in an Adriatic context. These infrastructures involved two types of linkage. The first involved the extension of railway and motorway networks, and the improvement of navigable internal waterways, in order to improve the accessibility of Porto Marghera with respect to the rest of the country and in particular to enhance the potential of the port as a gateway for the industrial developments taking place within the Po Valley area. The second concerned improvements to the intra-Venetian transport system, including a new road bridge and new car terminals, with a view to enhanced integration between the historic centre and the mainland. Despite the important effects of these changes on the spatial and functional structure of the historical centre of Venice, this area has never been completely integrated within the enlarged greater Venice region in terms of the levels of homogeneity and interchangeability typical of modern capitalistic industrial space. Bearing in mind the difficulty of overcoming the locational and environmental characteristics of the lagoon site and the difficulty of finding alternative locations for modern functions, the traditional lagoon city has tended to become an increasingly peripheral space in relation to the industrial zones of the nearby mainland. In fact, mobility problems and the advantages of mainland location have not only favoured the relocation of industrial activities but also of service industries such as banking and insurance which had previously been concentrated in the ancient artistic core of the city. From the 1970s onwards there has been a marked migration of service industries which tend to maintain only branch offices in the historical centre. The spatial separation of the city of the lagoon and the newer mainland port–industrial complex has therefore introduced a process of progressive functional impoverishment of the historic centre. Geographical polarisation brought about by the port–industrial development of the Venice area has increasingly emphasised the mainland zone. This has transformed Venice into a bi-polar conurbation, a characteristic which was fully recognised by the 1980s, by which time the mainland part of the Venetian commune was no longer regarded merely as an homogeneous industrial suburb.

Port–industrial development and the Venice protection issue

On the basis of optimistic forecasts of industrial expansion and traffic growth in the Venice area, the development of a third industrial zone was planned during the 1960s. This development plan foresaw the reclamation from the lagunar marshlands of 2940 hectares; 250 for the expansion of commercial activities, 484 for oil reception facilities, 327 for new equipment and services and 1879 for new industrial installations (Consorzio Obbligatorio, 1965). The planning authority took the view that these developments would strongly enhance the importance of the port of Venice in a Central European context. Mid-1960s estimates for 1980 envisaged a total general cargo throughput of 10 million tonnes (mt), industrial traffic amounting to 19 mt, and oil traffic of 50 mt in comparison with the 1965 totals of 3.5, 4.7 and 6.3 mt respectively. A basic objective was to serve an increasingly wide hinterland through more extensive and efficient transport infrastructures, so that Venice would act increasingly as a motive force in the industrialisation of the traditionally agricultural Veneto region. As the port authority pointed out,

> the General Plan fundamentally aims at bringing the rhythm of development of this unique and great Italian industrial port to the same intensity as other major industrial ports such as Rotterdam, Antwerp and Marseille . . . On the basis of the geo-economic situation of the Venice–Eugania region and the neighbouring areas, the preferred and most appropriate industries for the third industrial zone include engineering, iron and steel, aluminium, petrochemicals and fertilizer manufacture. . . With the new extensions in the industrial port of Marghera, Venice will have the largest industrial port in Italy, stretching over an area of about 4000 hectares.
> (Consorzio Obbligatorio, 1965, 17–21)

On 4 November 1966 Venice was flooded as a consequence of exceptionally high tides, a strong south-easterly gale and inadequate coastal protection. It may be argued that this dramatic event marked the end of port–industrial expansion within the Venice commune. Public concern led to widespread action aimed at the protection and conservation of the traditional historic core of the city. A UNESCO report published in 1969 stressed the demographic characteristics of the city, inhabited by an increasingly ageing population and by the accelerating migration to the mainland, and questioned the continuing integrity of Venice in the context of the Porto Marghera developments. New research centres were established during the early 1970s to study the environmental problems of the area. Private, mostly foreign, organisations took steps to finance the restoration of the fabric of the historic city. The Italian Government promulgated a 6th Special Law for Venice in 1973 outlining policies for the lagoon area as a whole. The idea of the third industrial zone was abandoned, and part of the already reclaimed land was flooded again. Some basic infrastructures such as the San Leonardo oil terminal were retained. Forecasts made in 1965 proved to be highly optimistic, as in 1979 general cargo traffic amounted to only 3.8 million tonnes, industrial traffic to 9.1 mt and oil traffic to 14.3 mt (Tonolo, 1989).

In the later 1990s the question of the protection of Venice has not yet been solved. Many investments have been made to reduce polluting emissions and waste in recent times (Rispoli, 1990). Nevertheless, the relationship between the environment and port–industrial development is still a thorny problem in the Venice region, both in the context of navigation and in relation to industrial accidents. Furthermore, Venice continues to be flooded by high tides which now occur about 40 times a year as compared with seven times a year at the beginning of the century (Zanda, 1991). Issues relating to the protection of Venice, in socio-political as well as environmental terms, now fundamentally condition the future evolution of the Porto Marghera port–industrial complex. The specific issues surrounding oil traffic are instructive. The 1973 special law for Venice provided, *inter alia*, for the diversion of oil traffic away from the lagoon. By the mid-1990s this law had not been put into effect. Oil traffic now represents about 45% of total port traffic and remains a key economic element in port throughput. Revenues derived from oil movements underpin the financial viability of the port, and crude oil and its derivatives unloaded in the port provide the raw material for the refining and petrochemical industries located in the port–industrial area. Projects aimed at connecting Porto Marghera industries by pipelines with other oil terminals on the northern Adriatic coast have recently been proposed. However, a lack of clear signals from the national government and the expectation of further job losses have combined to postpone a solution to this problem. The result is that the underlying controversy arising from conflicting factors involved in coastal zone management in this area is often rekindled, thus preventing continuity in the planning process.

THE PORT–INDUSTRIAL CRISIS AND THE ROLE OF NEW ECONOMIC ACTIVITIES

For Porto Marghera, as for other coastal industrial zones in advanced countries, the 1970s marked the end of a 20-year wave of continuous expansion. As far as port functions are concerned, despite good rail and road links with the hinterland and the widespread availability of container handling zones, the problems associated with the promotion of the necessary organisational changes in the management of port services including labour, together with the non-identification of marketing approaches consistent with the redefinition of inter-port competition as a result of the rise of intermodalism, have combined to prevent the port of Venice from playing a more dynamic role as a gateway for the regional manufacturing system. Venice still remains to a large extent an industrial port, with oil and industrial cargoes comprising 80% of total throughput. In the period 1980–92 commercial traffic grew from 12 to 20% of the total; but container and ro–ro traffics play a minor part in this sector (Provveditorato, 1993). In this context, the Padua dry port (about 20 km inland from Mestre) is maturing as the most attractive pole for the most advanced intermodal services within the regional transport chain (Zanetto, 1986).

As far as industrial functions are concerned, Porto Marghera has experienced extensive restructuring in recent years. The number of employees decreased from 30 000 to 29 000 during the 1970s, and fell further to 17 200 between 1981 and 1992 (Pugliese, 1993). About 50 hectares of industrial land have been abandoned as a result of the closure of the most obsolete basic industries. However, the penetration of service and research-based industries is still very slow, and Porto Marghera continues to be an area characterised by an overconcentration of heavy industries. Four largely state-owned major industrial types predominate – chemicals, metallurgy, shipbuilding and electricity – and employ 70% of the zone's labour force (Rispoli, 1990). The result is that the area's vitality depends to a large extent on European policy for basic industrial production, on public decisions regarding chemical and petrochemical industries, on state finance and, therefore, on the national economic situation as well as on the structural weaknesses of the national budget. In this situation, further reductions in the zone's labour force can be expected.

The crisis in the heavy industry subsector has reduced the role of the industrial complex in the growth of the Venice area and new economic activities have emerged as foci of spatial and economic organisation. In this respect two phenomena, the tourist industry and the development of the central Veneto region in a context of industrial decentralisation, deserve some attention. The lagoon city of Venice has emphasised its specialised tourist attractions since the 1960s, in the context of the wide range of tourism amenities available along the shores of the northern Adriatic. Today tourism is the most important economic activity in the historic city which attracts some 6 million visitors every year. Estimates drawn up in 1990 on the basis of economic development indicators related to tourist demand factors suggest that by the year 2000 the number of visitors will rise to between 7.8 and 8.6 million (Costa, 1990). However, about 80% of the number of visitors are day visitors (Costa and Van der Borg, 1993). In 1989, for example, of the 6.6 million people who visited Venice, 5.4 million (82%) were day visitors (or daily excursionists) and 1.2 million (18%) were residential tourists. About 65% of day visitors were Italian or foreign excursionists who stayed overnight outside the lagoon city but within a wide area of the mainland (extending to the provinces of Padua and Treviso) because of lower prices or a lack of city-centre accommodation (Costa and Van der Borg, 1993). Clearly, this phenomenon has a substantial impact on the physical fabric and on the socio-economic character of the city; and in the context of the central Veneto region the historical centre of the city has developed as one of the most important poles of attraction in terms of daily population movements (Costa, 1987, 1990).

The deep socio-economic crisis which has characterised Venice in recent years as a result of its inability to sustain further heavy industrial growth has coincided with remarkable socio-economic changes based on industrial decentralisation throughout adjacent areas of northern Italy, particularly in central Veneto. This has been derived from the exploitation of local economic, social and political resources, as various writers have recognised (Goodman et al., 1989; Conti and Julien, 1991). In other words, what was considered in the 1950s and 1960s as evidence of economic stagnation and inefficient productive systems (a lack of large industrial plants; supremacy of family-run firms and dominance of entrepreneurial figures in a context strongly characterised by rural socio-cultural values; the importance of local values such as collaborative culture, trust and social solidarity in the labour market) has proved to be one of the most important elements enabling many parts of the region to react positively to the recession of the 1970s. Already in the 1950s the Veneto experienced a first phase of diffuse industrialisation based on small family firms capable of competitive production thanks to low costs made possible by labour flexibility and widespread varied skills. In the 1970s the search for a maximum degree of productive decentralisation, up-grading technological innovations and market changes increased the ability of local systems in the region to compete on international markets. In the 1980s the economic vitality of many small and medium-sized industrial firms helped the Venice–Verona axis in particular to become a dynamic zone of economic activity (Soriani, 1993).

These changes in the industrial geography of the Venice hinterland, together with the wider trends towards counterurbanisation which have affected the Veneto as a whole during the 1970s and 1980s, have produced an increasingly urbanised countryside focused upon Verona and upon the polynuclear urban region centred upon Padua, Treviso and Venice–Mestre. In this increasingly structuralised context, despite the plurality of foci, the Venice mainland port–industrial zone has gradually lost its homogeneous industrial character. The continuous outflow of population from the historical centre of Venice towards the small municipalities in the outer ring, and the birth of new small-scale entrepreneurial initiatives, have therefore helped Mestre (in competition with Padua) to mature as a service centre for population and business enterprises since the 1970s (Costa, 1987). As a consequence, spatial and functional relationships between the three main poles of the Venice urban system – the historic core, Mestre and Porto Marghera – have radically changed. The lagoon city and Mestre have turned out to be the most important centres of employment and services within the Venice urban system, while Porto Marghera has increasingly become a negative element.

The steadily decreasing role of Porto Marghera in the continuing evolution of the Venice area, the rise of new economic specialisms on the mainland, the impressive development of tourism and the issues surrounding the protection of the old city, have combined to underline the perception of Porto Marghera as an element upon which the economic destiny of the region as a whole is decreasingly dependent. Porto Marghera is considered as a problematic inheritance from an industrialisation process that has left behind many hard social and environmental problems in the Venice region. In this context, port and industrial authorities have begun to speak of the need for the protection of Porto Marghera (as opposed to that of Venice) in the current socio-political and economic climate.

PORTO MARGHERA IN A CONTEXT OF REGIONAL CHANGE

The development of Porto Marghera initially enhanced the potential value of north Adriatic industrial sites in relation to the core zone of the modern Italian economy centred on Milan. Industrial location on the mainland facilitated the processing of raw materials imported through the port, with good transport links with the most dynamic development zones in the country and with no serious problems concerning land sites or infrastructural provision. By restructuring its urban waterfronts and creating a new industrial port, Venice emerged increasingly as a gateway for raw materials within the restructuring space economy of northern Italy (Zanetto, 1985).

The development of the second industrial zone at Porto Marghera reflected trends apparent in similar coastal port–industrial contexts in other West European countries and in Japan (Hoyle and Pinder, 1981; Vallega, 1992). The implementation of this development, however, and the proposal for a third industrial zone, is related to the wider context of the debate about the problem of promoting large-scale industrialisation in areas considered to be peripheral to mainstream Italian economic growth. Veneto was long considered as a kind of *mezzogiorno del nord* (a southern Italy of the north) because of its agriculture-based economic structure and because of emigration. It was expected that the limited network of small-scale family-run industrial activities in the Veneto, regarded as a legacy from Italy's late modernisation, would largely be eliminated by the development of large industrial plants. Porto Marghera, and especially the projected third industrial zone, were regarded by entrepreneurs and regional planning authorities as a basic step essential to the enhancement of the effects of the growth pole at the regional level and as a way of attempting to reduce the substantial gap between Veneto and other north Italian areas in terms of large-scale industrialisation (Consorzio Obbligatorio, 1965). Just prior to the 1970s recession, however, the environmental problems of the lagoon and the issues surrounding the protection of Venice led to the abandonment of the third industrial zone project and thus effectively eliminated any opportunity to verify these ideas.

Today, Porto Marghera remains a coastal industrial zone characterised by the typical features of a first-generation MIDA: raw materials and oil traffics are still the core of port functions, basic industries are the prevalent activities in the industrial zone and the penetration of more highly integrated activities in the urban area has been slow. Despite the role that Porto Marghera still plays in the local labour market, it lacks integration within central Veneto, the development of which in recent decades has often been cited as a clear illustration of the characteristics of the affluent Italian periphery. It would, however, be wrong to regard Porto Marghera solely as a legacy from an industrial experience that has bequeathed hard social and economic problems in the Venice area. The crises affecting the industrial functions on which Porto Marghera has built its role as a gateway since the beginning of this century, together with economic restructuring involving higher-order functions in central Veneto (Zanetto and Lando, 1991; Soriani, 1993), appear to suggest that Porto Marghera might play a more dynamic role in the future.

By encouraging the introduction of new functions including research-based manufacturing industries more closely integrated with the regional system, a higher order of urban services, and a more sophisticated regional distribution system based in transport logistics, Porto Marghera could develop as a basic resource for the metropolitan development of central Veneto. In theory, the idea of developing a more specialised role for Porto Marghera in these contexts rests on the ready availability of land for development or redevelopment and on the well-established transport infrastructures already in existence. In the context of demand for land for industrial development in central Veneto, it should be noted that counterurbanisation, polycentrism and industrial diffusion have produced serious mobility problems, environmental impacts and land-use conflicts.

In reality, the possibility of integrating Porto Marghera within a wider metropolitan development scenario depends on the ability of the local area to promote the transition towards more complex forms of production and patterns of settlement. Within the port area itself, increased functional specialisation in selected sectors and the promotion of enhanced cooperation with the major nodes of the regional distribution system (especially the Padua inland port) require substantial improvements in several important areas. These include port management, labour organisation, a competitive approach to marketing port services, and the revision of port layout. Some steps in these directions are already being taken. The port authority has delimited a zone within the commercial sector of Porto Marghera where a free trade zone (currently located in the Marittima port zone) should be relocated. This decision may represent the starting point in a process leading to a new layout of port facilities at Porto Marghera involving the concentration of commercial traffic in the commercial zone of the port where intermodal facilities are available, and the development of a new passenger terminal in the Marittima sector. As for port management and organisation, a new national law introduced in 1994 provides for improvements in this context which *inter alia* may start the complex process leading to the adoption of more vigorously market-orientated approaches in port management.

In order to attract and promote the development of new high quality manufacturing and service industries at Porto Marghera, a technopole (similar to those located near Verona and Padua) is in the process of being established in the first industrial zone, that is in the area which suffered most from the restructuring processes of recent decades. This technopole will include, in addition to research activities related to industrial processes and materials and to environmental and marine technology, a business innovation centre of great importance in helping small firms to face the problems of technological innovation and high quality services (Soriani, 1993). Obviously it is not yet possible to say to what extent these innovations will succeed. However, there can be little doubt that the quality of the functions performed by Porto Marghera will not be enhanced by these initiatives alone. What is lacking is a clear vision of the future role and character of Porto Marghera with which all the agents involved in the transition process towards a more complex and integrated pattern of production and settlement can identify and can support. At present, such a coherent vision does not exist, and too many institutional barriers and conflicting expectations continue to prevent its emergence.

ACKNOWLEDGEMENTS

I would like to thank Professor G. Zanetto (University of Venice) and Dr R. Bruttomesso (University Institute of Architecture, Venice, and International Centre Cities on Water, Venice) for their helpful suggestions and indications on the topic, but they bear no responsibility for opinions or conclusions stated herein.

REFERENCES

Carbognin, L., Frankenfield Zanin, J. and Ramasco, C. (ed.) (1993), *A guidebook on the lagoon of Venice. Environment and problems* (Venezia: ISDGM–CNR).

Chinello, C. (1979), *Porto Marghera 1902–1926: alle origini del problema di Venezia* (Venezia: Marsilio).

Consorzio Obbligatorio per il Nuovo Ampliamento del Porto e della Zona Industriale di Venezia–Marghera (1965), *The development area of the commercial and industrial port of Venice* (Venezia: Fantoni).

Conti, S. and Julien, P.A. (ed.) (1991), *Miti e realtà del modello italiano* (Bologna: Patron).

Costa, P. (1987), 'Le trasformazioni recenti dell'economia del sistema urbano veneziano', in IReR/Progetto Milano and Fondazione Agnelli (eds), *Il sistema metropolitano italiano* (Milano: Franco Angeli), 157–80.

Costa, P. (1990), 'Il turismo a Venezia e l'ipotesi Venetiaexpo 2000', *Politica del Turismo* 1, 1–17.

Costa, P. and Van der Borg, J. (1993), 'The management of tourism in cities of art', *Quaderni CISET* 2 (Venezia: Università degli Studi di Venezia and International Centre of Studies on the Tourist Economy).

Costa, P., Dolcetta, B. and Toniolo, G. (1971), 'The new scale of the city', *The Architectural Review* 891, 310–12.

Ente della Zona Industriale di Porto Marghera (1990), *Porto Marghera, Venezia e l'ambiente lagunare* (Treviso: Dosson).

Goodman, E. and Bamford, J. with Saynor, P. (eds) (1989), *Small firms and industrial districts in Italy* (London: Routledge).

Hoyle, B.S. and Pinder, D.A. (eds) (1981), *Cityport industrialization and regional development: spatial analysis and planning strategies* (Oxford: Pergamon).

Hoyle, B.S., Pinder, D.A. and Husain, M.S. (eds) (1988), *Revitalising the waterfront: international dimensions of dockland redevelopment* (London: Belhaven).

Provveditorato al Porto di Venezia (1993), *Statistiche del traffico* (Venezia: Ufficio Statistica Provveditorato al Porto di Venezia).

Pugliese, T. (1993), 'Porto Marghera nell'economia della città metropolitana', *Oltre il ponte* 43–44, 58–71.

Rispoli, M. (1990), 'Porto Marghera, una zona industriale in evoluzione', in COSES and Comune di Venezia (eds), *Porto Marghera. Proposte per un futuro possibile* (Milano: Franco Angeli), 13–48.

Soriani, S. (1993), 'Problems and opportunities for future development: new prospects for the Veneto region', in Horvath, G. (ed.), *Development strategies in the Alpine–Adriatic region* (Pécs: Centre for Regional Studies), 109–41.

Toniolo, G. (1972), 'Cento anni di economia portuale a Venezia', in Costa, P. (ed.), *Il porto nell'economia veneziana* (Venezia: COSES), 33–74.

Tonolo, C. (ed.) (1989), *Il porto di Venezia* (Venezia: Marsilio).

Vallega, A. (1992), *The changing waterfront in coastal area management* (Milano: Franco Angeli).

Zanda, L. (1991), 'The case of Venice', in Frassetto, R. (ed.), *Impact of sea level rise on cities and regions* (Venezia: Marsilio), 51–9.

Zanetto, G. (1985), 'Primo Lanzoni, ovvero l'economia come antitesi all'ambientalismo nel pensiero geografico ottocentesco', *Ricerche Economiche* **XXXIX**-1, 70–103.

Zanetto, G. (1986), 'Regionalizzazione costiera e Porto Marghera', in CNR (ed.), *L'umanizzazione del mare. Riflessioni geografiche sugli spazi funzionali costieri* (Roma: CNR), 272–82.

Zanetto, G. (1989), 'Le ragioni economiche-geografiche dell'industrializzazione costiera e la loro dinamica recente', *Terra* **50**, 30–4.

Zanetto, G. (1992), 'Suburban growth in the Venetian Central Metropolitan Area', in Lando, F. (ed.), *Urban and rural geography* (Venezia: Cafoscarina), 141–7.

Zanetto, G. and Lando, F. (1991), 'La dinamica territoriale dell'industria veneta', in Bernardi, R., Zanetto, G. and Zunica, M. (eds), *Il Veneto. Diversità e omogeneità di una regione* (Bologna: Patron), 11–58.

14 Combined Transport in Italy: The Case of the Quadrante Europa, Verona

CLAUDIA ROBIGLIO
Institute of Economic Geography, Faculty of Economics, University of Verona, Italy

During recent decades, a series of factors of a diverse nature have caused profound changes in medium- and long-distance transport systems. In demand-led economies, the characteristics of manufacturing industry have been substantially modified during the 1970s and 1980s. Earlier structures, made up of large companies, vertically integrated and operating in national and international markets, together with small firms operating in purely local markets, have given way to an economically segmented structure made up of large companies connected in various ways to medium- and small-size firms and small companies that operate in market niches. During the 1980s there was a widespread movement from the older towards the newer structures throughout industrial zones and urban and non-urban industrial systems in the advanced countries. In Western Europe, Italy has followed (and in some senses led) this process: in particular, industrial complexes of relatively small size have come to involve a variety of firms interacting in various ways and constituting areas of specialisation identifiable in regional terms. The literature on this subject is now substantial (e.g. Pyke *et al.*, 1990; Conti and Julien, 1991).

In this context this chapter considers, first, some aspects of the transportation response to the new productive systems, outlining the logic that connects post-Fordism, neo-Fordism and flexible production systems with multimodalism and intermodalism in transport terms. After establishing the European framework of these developments, the chapter considers the more specific issues associated with road–rail intermodality in Italy, as illustrated by the specific case of the Interporto Quadrante Europa, an intermodal freight centre at Verona. The essential questions raised by this discussion concern the impacts upon traditional coastal ports (and the coastal zones within which they are located) of the new multimodal and intermodal environments in which coastal zones and ocean ports may appear to be involved in a process of relegation to the periphery of economic space.

Cityports, Coastal Zones and Regional Change. Edited by Brian Hoyle.
© 1996 John Wiley & Sons Ltd.

PRODUCTIVE SYSTEMS AND TRANSPORT NEEDS

The debate about the passage from Fordism to post-Fordism and the present characteristics of the coexistence of Fordism, neo-Fordism and flexible production is still open (Dicken and Lloyd, 1990; Gertler, 1992; Storper and Scott, 1992; Knudsen, 1994). However, it is certain that the present systems and modes of production, the variety of demand and supply of goods both for investment and consumption, and the globalisation factor in the context of plant location, suppliers and markets, have together involved a remarkable growth in freight transport. Moreover, movements that increasingly involve light industrial products (semi-manufactured goods, component parts and finished products), and medium- and long-distance transport, require solutions to the problem of providing transport on as rational and inexpensive a basis as possible.

Complexities which today characterise the transport sector by reason of its growing integration with productive systems will undoubtedly increase in the future. In this context of considerable (and growing) structural productive capacity, transport can no longer be considered with regard to regional economic growth as an independent variable, but rather as one of the essential elements of territorial organisation. This integration in turn depends on the conceptual revolution now taking place within the productive system. Transport change constitutes an important element within the whole logistic chain upon which any modern enterprise strategically depends in order to respond to the challenges of market specialisation and globalisation.

The most pressing requirements of modern transport, cheapness and rapidity, together with other factors such as security and environmental concerns, have produced a radical change in approaches to different transport modes (rail, road, sea, air and inland waterways): formerly considered as largely self-contained, individual modes are now regarded as interdependent elements within an interactive system. In other words, it is no longer a matter of considering the use of just one single mode in the organisation of transport from producer to consumer, but of taking advantage, as far as possible in a coordinated way and through a single operator, of several modes organised as an interdependent multimodal chain, synchronising the different operations involved in an intermodal environment (Vallega, 1984, 1985; Hayuth, 1992).

Load unitisation has represented a first step towards transport rationalisation. Bulk breaks have been reduced to a minimum by forwarding different consignments with the same destination as a single module (container, semi-trailer or swapbody). Hence goods are consolidated according to geographical destination, and thus become simply moving unit loads for most of the journeys involved (Rizzo, 1980). The logic of multimodality – road–rail, road–rail–sea, road–waterway and other combinations – has produced a rapid improvement in the quality of transport services and a rapid advancement in the application of the rationale of the concept of unitisation: from the consolidation of goods into load units to the inception and generalised take-over of unit loads (Vallega, 1984,

1985; McKenzie *et al.*, 1989).

All this has involved considerable spatial change in the organisation of coastal and inland transport nodes. In fact, the process of concentration of intermodal load units with a single destination has triggered a selection process among the nodes themselves. Some have acquired dominant functions – for example, as loading port centres or as inland loading centres (Hayuth, 1982; Slack, 1990) – and this has produced a close connection, almost a continuum, with the most important road and rail traffic routes, originating landbridges, mini- and micro-landbridges and basic links between inland nodes. Thus, in a context of intermodality, substantial changes occur in the spatial and functional relationships between ports, hinterlands and the oceans, and also between ports and cities (Hayuth, 1988; Hoyle, 1988; Vallega, 1992). Moreover, as a result of the widespread use of containers, coastal nodes have acquired new functions in the context of load unit transportation and have adopted new and expensive equipment and sophisticated intermodal techniques. In physical planning terms, a need has emerged for substantial areas to be set aside for marshalling and storing loads, and for new railway terminals within or immediately adjacent to ports and at inland locations.

The decision-making process involved in multimodal and intermodal transport has increasingly emphasised the critical role of the inland nodes at the expense of the coastal ports which are increasingly perceived as transit points within an intermodal system. It is increasingly at the inland nodes that the industrial and commercial activities and the operational and managerial headquarters of the various transport operators are located. Intermodal railway terminals acquire a wide range of services, according to their relative importance and that of their catchment area. Therefore we have structures that range from mere transhipment terminals to more complex freight villages where a wide range of services for goods, transport equipment, modules and staff engaged in the different activities are concentrated (Table 14.1).

These processes involve the establishment of connections between maritime nodes, inland nodes and the transport networks and systems involved. These systems operate at different territorial scales reflecting the characteristics of their component elements. However, it is possible to recognise, at least in embryo and generally speaking, the following types:

● regional systems in which links are established between a dominant intermodal port, feeder ports, basic transhipment terminals and possibly a freight village;
● national systems including higher level intermodal centres and freight villages;
● international systems in which all ports, intermodal centres and freight villages operate at a high level.

This is an oversimplification, of course, because some elements might operate within different systems that consequently overlap and interact.

Table 14.1. Typology of intermodal facilities

Intermodal structures	Main functions and activities			
Terminal	ITU exchange between carriers			
Goods centre	ITU exchange between carriers	Loading and warehouse consolidation		
General warehouse	ITU exchange between carriers	Loading and warehouse consolidation	Stocks and customs, fiscal and financial services	
Freight village	ITU exchange between carriers	Loading and warehouse consolidation	Stocks and customs, fiscal and financial services	Management services, information and freight exchange

ITU = Intermodal transport unit (containers, swapbodies, trailers, etc.).

Table 14.2. Combined traffic in Europe and in Italy ('000 shipments)

	1988	1989	1990	1991	1992	1993
Europe						
International	357	422	543	638	640	623
Domestic	654	578	640	588	649	674
Total	1011	1000	1183	1226	1289	1297
Italy						
International	186	214	254	283	300	335
National	53	59	73	87	102	113
Total	239	273	327	370	402	448

Source: A.T. Kearney/UTRR

INTERNAL INTERMODALITY WITHIN THE EUROPEAN UNION

The particular emphasis placed upon internal intermodality by the European Union (EU) testifies to the growing importance that this transport phenomenon is acquiring. At the Community level and at national levels within member countries, transport policies in the 1990s have been strongly orientated towards the creation of a trans-European network of combined transport, whose initial and final road sectors are as short as possible and the longest sectors are by rail, by inland waterway and by sea (Figure 14.1). As internal intermodality systems consolidate and develop within this European network of combined transport, there is a high level of dependence in practical terms on swapbodies and semitrailers (see Table 14.3).

In October 1993, the EU identified those rail connections in need of adjustment and fixed a timetable for carrying out the necessary work, while also drawing attention to related projects involving transhipment facilities and the availability of suitable rolling stock. There are 37 connections expected in the first phase, nine of which are in Italy (two involving Verona) and 20 in the second phase, five of which are in Italy. A new transport geography is emerging within this European system, based on intermodality, and the inland intermodal centres constitute the keystone of the whole system. These centres are partly identifiable according to their catchment areas, but also by the specialised roles they perform.

The precise characteristics and role of each inland intermodal centre are shaped by the final destinations (national or international) that are served, by the specific main transport links utilised and by the different kinds of loading units used. Inland intermodal centres respond and develop as the pattern of traffics served consolidates – some specialising in long-haul traffic between inland destinations, for example, while others deal particularly with traffics originating in or destined for overseas destinations. Some specialise in the formation of trains whose composition is blocked but which accept mainly semitrailers and swapbodies (the latter are less robust and therefore inappropriate for maritime transport but, being more capacious and manageable, are more suitable for land transport). In fact, inland nodes have to satisfy the diverse requirements of the load units, since a consignment may comprise one single land sector, one land and one maritime sector, or some other combination of sectors. In one case we might, for example, have very short movements at the terminal that involve vertically shifting loads from a road vehicle to a rail carrier and vice versa. In fact, the departing block train, sometimes operating with just a daily frequency, is the critically important rendezvous for the load unit. Precise timing is essential, especially for road vehicles, as parked lorries or trucks (like berthed vessels in port) represent an unprofitable cost. Where the rendezvous is with a vessel in port, however, departures are less frequent and therefore the load unit might have to be parked for some time. All intermodal connections, however long or short the time involved, necessarily involve the use of additional facilities that consume space, provide equipment and services and fulfil a variety of functions.

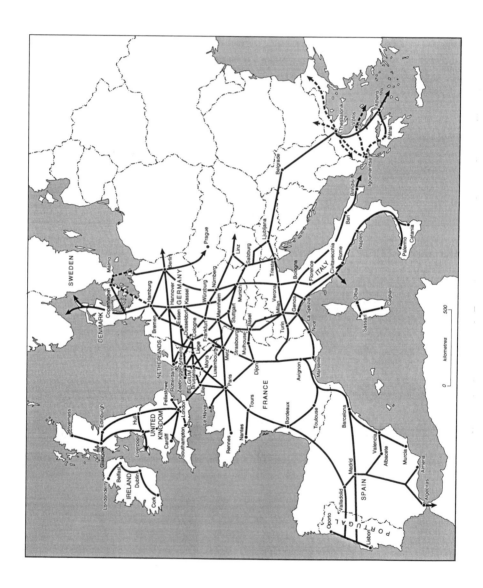

At present the expansion and development of intermodal transport systems within Europe suggests that road–rail combined transport is reducing the importance of sea transport between European countries. However, a knowledge of what is happening, and might happen, to the European network of combined transport allows ports either to pursue the development of internal intermodality, where this is already established, and/or to identify further possibilities to develop prolongations of land links. One possible example in this context is the planned Greece–Italy–Northern Europe multimodal route, designed to channel all multimodal traffic with an origin or destination in Greece through the port of Brindisi. This is an example of an EU-funded pilot scheme, an illustration of the PACT (Pilot Actions for Combined Transport). Some questions are raised, however, about the balance between internal combined transport and external maritime traffic, and the eventual use of ports solely for extra-regional connections. Port nodes must decide whether such changes are due to infrastructural and/or organisational deficiencies and whether regaining lost traffic is therefore possible; or whether operating processes are such, at least in the present stage, as to make it impossible to achieve this kind of synergy. In this latter context, coastal nodes may ultimately have to pursue other functions.

ROAD/RAIL INTERMODALITY IN ITALY

Long-haul trucking dominates land transport systems in Europe. The imbalance in the modal share of goods transport at the European level (72% road, 18% rail in 1991) is particularly serious in Italy (82% road, 13% rail). Due partly to the greater flexibility of road transport but mainly to political decisions taken in the past decade, this imbalance involves a series of penalties which are no longer acceptable: increased transport costs (energy and congestion of the road network), the costs of wear and tear on the infrastructure and the increased impact on the environment. At a community level and even more at a national level (because of the structure of the Italian productive system, and the country's geographical shape) the readjustment of policies is also essential to the intensification of economic relations between the EU and the countries of Eastern Europe. Readjustments towards a new balance cannot but rely on the main nodes of the rail network, and in particular on those which are more closely connected with other transport nodes and are located in areas of increased productive capacity. A significant stage in the development of road–railway intermodality was reached when the percentage of goods trains in the EU specialising in intermodal traffic increased from 9% in 1987 to 24% in 1993.

If a particular intermodal centre is to become an important element in the development of the local economy (as a decisive factor affecting localised entrepreneurial choice, and as an opportunity in terms of the containment of transport costs), it needs to concern itself with the total network of intermodal

Figure 14.1. (*opposite*) European corridors of combined transport

centres to understand the impact on economic space and its organisation. This requires a consideration of spatial distribution, different efficiency levels, specialisation and possible network relations.

Table 14.3 provides a typology of intermodal facilities available within the road–rail system in Italy. The relative significance of the intermodal centres is summarised and they are distinguished according to the organisation which owns them and the type of loading unit movement involved. As one can see, the clear predominance of the private sector is achieved only in single-container movement, evidently a situation inherited from the strong development of maritime containerisation in the 1960s and from the fact that often such private sector depots merely store empty loading units; while a wider range is found in the public sector. Another observation worth making is that with 42 intermodal facilities in a position to operate combined transport (transferral of swapbodies and semitrailers for land movement), Italy finds itself in this context among the best-placed countries in Europe.

This is important when we consider that combined traffic is the mode with the highest rate of growth in moving goods by railway in recent years. We may take into account the prospect of an even greater increase in the next few years if Austria is also forced to adopt restrictions on trans-Alpine road transport flows similar to those applied in Switzerland. Also the geographical distribution of the intermodal facilities is of some interest since the whole of the private sector as well as half of the public sector is found in northern Italy and Tuscany (in this case mainly linked to container movement in the ports of Livorno and La Spezia), in evident rapport with the major economic–productive ability of these regions.

FREIGHT VILLAGES

We now focus on those intermodal centres known as freight villages in which, as well as road–rail intermodality, there is also a concentration of structures and

Table 14.3.   Intermodal centres in Italy

	Loading units	State Railway	Private	Total
Combined traffic	Containers Swapbodies Trailers	27	13	40
	Swapbodies Trailers	1	1	2
	Total	28	14	42
	Containers	9	39	48
	Overall	37	53	90

Source: Ferrovie dello Stato (FS) (State Railway) — *The Supply of Goods 1993/94.*

logistic services for freight and operators (Figure 14.2). In the early 1990s the Italian national general transport authority foresaw the development of nine level 1 freight villages together with a further 16 level 2 freight villages to which, eventually, another 12 would be added. The level 1 and level 2 freight villages were planned within the framework of a coherent national transport system. However, while the level 1 freight villages are confined to the economically strong northern regions of Italy and Tuscany (excepting that envisaged in the Naples area), three-quarters of those at level 2 are foreseen in the central–southern area, to counterbalance the northern bias in level 1. This reveals at least

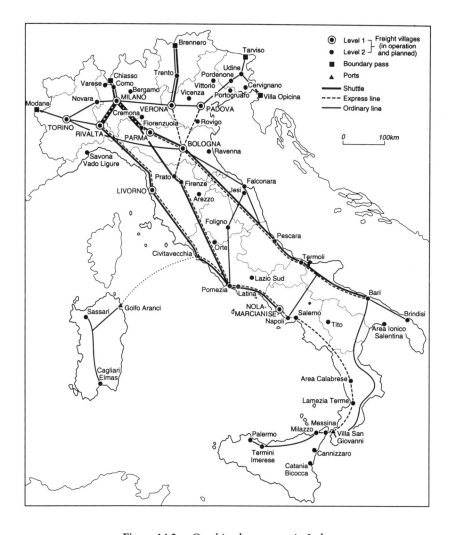

Figure 14.2. Combined transport in Italy

an implied capacity of freight villages to assist and perhaps even to promote development.

Limited resources and the present national and international economic climate have caused the Italian Government more recently to abandon the distinctions between level 1 and 2 freight villages in the name of regional balance. This (and a certain lack of initiative and support at the local level) will probably reshape the original design. At present only seven out of the nine major freight villages are active, connecting with the principal multimodal corridors and in particular with their nodes, namely those of Torino-Orbassano (Turin), Rivalta-Scrivia (Alessandria), Parma-Fontevivo, Bologna, Padova (Padua), Verona and Nola-Marcianise (Naples). All except the last-named one are located in northern Italy, and the three freight villages of Bologna, Padua and Verona were already being planned in the 1970s, located in two regions that in many respects form part of the 'Third Italy', with a dynamic industrial structure and an export-based economy. The seven operating major freight villages, and some others from level 2 already operating or about to open, put Italy in an extremely advantageous position in the context of European markets, not only because of their infrastructure but also because of their management systems and services which involve the collaboration of several public and private organisations within a programme supported by the state.

The relative importance of the operating level 1 freight villages in terms of throughput is shown in Table 14.4, which also displays a process of specialisation. Turin (Torino-Orbassano) operates mainly as an organiser of combined national traffic (in part brought about by FIAT); Parma, which began activity in 1991, because of its geographical position, is developing a role organising the Apennine area towards the ports of La Spezia and Livorno, and will probably become increasingly important for container traffic. Bologna shows a marked specialisation in domestic combined traffic – operating overall as a forwarding point to and from the ports. Padua (Padova) serves Friuli-Venezia Giulia and part of Veneto; it is the most important freight village for export

Table 14.4.   Flow of intermodal loads through Italian freight villages, 1993

Freight village	Containers	Combined transport international	Combined transport domestic	Internal movements	Total
Bologna	2 900	3 700	29 500	1 900	38 000
Padua	39 700	1 400	14 900	6 400	62 400
Parma	0	0	3 000	0	3 000
Turin	2 600	2 200	4 400	8 700	17 900
Verona	15 900	124 200	4 500	300	144 900
Total	61 100	131 500	56 300	17 300	266 200

Source: Statistics CEMAT (Milan).

container traffic and also plays a remarkable role for domestic traffic moving in a southerly direction. Verona functions largely as an international import/export centre on the Brenner axis and is especially important for international combined traffic with around 95% of the total international combined traffic of the five Italian freight villages shown in the table. Verona currently handles few containers. The other two, Rivalta-Scrivia and Nola-Marcianise only recently became operational and do not yet show any clear specialisation.

THE FREIGHT VILLAGE OF VERONA

Freight villages are now well developed in Italy. The example of Verona (Figures 14.3 and 14.4) is an interesting one to look at in more detail, on the one hand because it has acquired a remarkable significance at an international level in the European network of combined transport, and on the other hand because it illustrates a range of influences on the economy and on urban morphology, at the local and regional levels.

Verona's nodal position, at the cross-roads of two of the major traffic routes of Padania, has always been a precious natural inheritance for the city, determining its political and economic history as well as its urban evolution. In the past, Verona was largely a commercial city with an economy predominantly based on agriculture and on the military services. Some industrialisation took place in the city and its environs between the two world wars but this developed significantly only after 1945. Today the production system seems well balanced in its essential elements; industry and agriculture are well integrated with one another and with other sectors such as commerce, trade and tourism. Verona is, however, only one of a number of poles of industrial concentration within a much larger area, characterised by the spread of industry, by inter-sectoral diversification and by a marked inclination towards export-orientated production. These elements, together with the helpful creation of an Industrial Agricultural Zone (ZAI) (an area to the south of the historic city exceeding three million square metres where industries of diverse kinds are concentrated) created by the 1970s a city of trade and transport.

The development of the Verona freight village is based on the idea that, in place of the traditional geographical nodality it would be advantageous to establish a 'constructed' nodality in the knowledge that established advantages would eventually probably decline, because of changes in society and the economy in general and in the transport chain in particular. Near Verona, at the intersection between the Prealpino Padano multimodal corridor on which the east–west flow runs and the Central-Ridge which gathers north–south traffic, a planned structure of transport facilities and services has begun to take shape. In 1988 the Quadrante Europa or European Quadrant was born and in an area of over 2.2×10^6 m^2 (the completed project will eventually occupy 5×10^6 m^2) is already active. The rail terminal for combined traffic, the general warehouses, the customs, the forwarding centre, the trucking centre, the back-up services, the

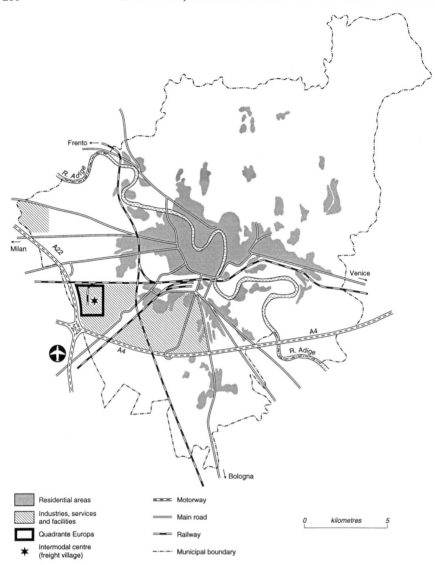

▨ Residential areas	▭▭ Motorway
▨ Industries, services and facilities	═══ Main road
☐ Quadrante Europa	▰▰ Railway
★ Intermodal centre (freight village)	─·─·─ Municipal boundary

0   kilometres   5

Figure 14.3.   Verona, Italy: the location of the Quadrante Europa

parking zones and the management centre are all operational: this part of the European Quadrant constitutes the freight village (Figures 14.4 to 14.8).

The core of the freight village is the railway terminal. This is considered one of the most important on the continent in terms of operating capacity and efficiency. It covers an area of $350\,000$ m^2 (potentially $800\,000$), utilises 12 tracks (with a further 17 in support) and is equipped with lifting gear which enables movement of over $100\,000$ intermodal transport units annually ($164\,000$ in 1995).

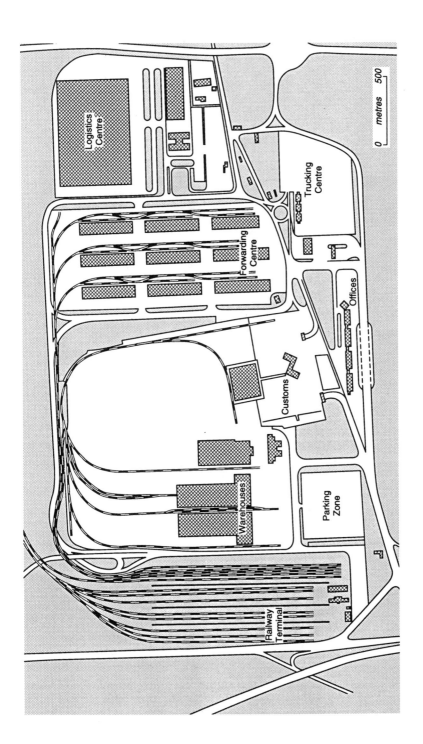

Figure 14.4. Quadrante Europa: the Verona freight village

Figure 14.5. The Quadrante Europa inland multimodal terminal, Verona, Italy. (Consorzio ZAI)

The terminal area accommodates the railway offices, railway police, rail customs, customs officers, veterinary office, the terminal's management company (CEMAT) and various private operators. In 1993 around 20 connections were made (half of the total number made by all the Italian freight villages), connecting Verona with the most important industrial activity of Central and Northern Europe, and recently also with the Eastern European countries, by unit train loads and shuttles.

It is estimated that the area of gravitation to the Veronese freight village has a radius of 100–150 km and therefore comprises, as well as the province of Verona, areas of considerable economic weight and productive capacity within the neighbouring provinces of Vicenza and Brescia, extending as far as Milan, Mantova and Trento. The proximity (1 km) of the city airport, and of the river port of Mantova, are significant advantages that enhance the important role which the freight village of Verona performs in the Italian intermodal system as a whole.

The management centre houses, as well as the company providing administration and maintenance support to the freight village facilities (Consorzio ZAI), the merchandising branch of the state railway (FS), the customs chemical laboratory, operators of different sectors (banks, insurance

Figure 14.6. The Quadrante Europa inland multimodal terminal, Verona, Italy, under construction. (Consorzio ZAI)

companies, forwarding agents, transporters, consigners, warehouse management, back-up services), telecommunications offices, the telematics link between supply and demand of transport (TRAMITE), as well as the headquarters of the European School of Logistics, a research centre of the National Research Council (CNR), and the consortium of the universities of Padua, Verona, Rome and Calabria for the study of industrial and managerial economics (CUEIM).

The General Warehouses, once a simple storage point, have become a logistic centre taking on various functions: collection and distribution of goods for a third party, finishing operations, quick-freezing, bond warehousing, packing and storage of grain products, customs services, and all the container operations and groupage (arranging links with the rail network). Also important are the customs enclosures for the warehousing of motor vehicles from foreign car companies, which have chosen Verona as the central point for the entire national territory: the parking capacity is for around 10 000 vehicles (Figure 14.4).

SPATIAL IMPACTS

It is worth highlighting at this point the spatial changes which the presence of the freight village have caused in the European Quadrant itself and in the city of

Figure 14.7. The Quadrante Europa inland multimodal terminal, Verona, Italy: layout and facilities under construction. (Consorzio ZAI)

Verona. First, the change in usage of an area previously consisting predominantly of agriculture with some specialised cultivation. This change of usage, apart from all the legal problems connected to the expropriation of land areas, has involved the development of a road network infrastructure to link with the motorway, a new rail connection with the principal station of Verona, and a rail infrastructure now serving the city and also operating as a terminal and as the general warehouse and headquarters of the forwarding agents.

Second, the European Quadrant has brought about a concentration of businesses; there are about a hundred offices installed in the European Quadrant and around 1500 employees now work there in this sector. But this concentration was only partly attracted to the Veronese node by the presence of the inland load centre. Verona has an established tradition of exchange and commerce and its importance to the transport system long preceded the development of the European Quadrant, but undoubtedly the majority of the 3500 workers of the transport sector (transporters, consigners and customs workers) operating in 1989 worked within the various public structures and offices present in the Quadrant. Today, mergers and firms ceasing activity due to the liberalisation of the traffic in the Community, and the strong competition derived from this, have reduced the numbers. Even now, however, one can estimate at around 2500 units

Figure 14.8. The Quadrante Europa inland multimodal terminal, Verona, Italy: layout and facilities. (Consorzio ZAI)

the businesses in operation and of these the majority are national operations, with international connections.

Besides this polarisation of transport sector enterprises, in the Quadrant and in the Veronese node at large, the freight village has attracted, by its location and structure, the establishment of the Autogerma plant (the main foreign motor importers in Italy which handled over 300 000 vehicles imported in 1993). The company has built its headquarters, with the spare-parts warehouse for the whole national territory, in close proximity to the general warehouse whose customs enclosure and some installations it uses. Looking toward the future, the construction of a wholesale centre is foreseen, including the relocation of the city's fruit and vegetable wholesale market, one of the most important at the national level overall for the role it plays as an exporter. Hence, a synergetic continuity between transport and trade, between large commercial distribution systems and major transport networks, is being achieved.

The impact of the European Quadrant on the urban patterns and functions of Verona is also of interest. The convenience of relocating to the Quadrant in order to take advantage of the possible synergy between different operators has brought about the transfer of a certain number of enterprises operating in the transport sector. Few in truth have abandoned the offices and the areas they occupied in the city and the vast service and industrial zone to the south of the historical centre. Generally speaking there has been some duplication of offices, and in most cases their relocation has had only a local importance. More importantly, a need has arisen – due on the one hand to the freight village and the volume of traffic it has caused, and on the other hand to the industrial development of the entire area which has Verona as its centre – to provide the Quadrant with an adequate structure of warehouses and depots. The logistic evolution of the industrial enterprises and the spread of 'just-in-time' methods – which involve a drastic reduction in storage of supplies and products close to industries, together with an externalisation of these functions and others connected to it and, ultimately, the use of transport modes as travelling warehouses – has meant the introduction of special warehouse facilities by industrial and commercial firms and has pushed the major consignors to offer this service in their premises side by side with traditional groupage.

The development of the Quadrante Europa and the associated restructuring of industrial and transport activities in the greater Veronese region have involved expansion, relocation and the utilisation of new areas (especially in the Zona Agricola-Industriale (ZAI)), the modernisation of buildings and the refining of transit and transport techniques. But undoubtedly a major urban impact has been that brought about by the transfer to the European Quadrant of the customs and the general warehouses, and further major changes will be caused in the future by the transfer of the fruit and vegetable market. The migration of the customs, formerly located in the historic core zone and originally associated with river traffic on the River Adige, has not had the effect of freeing substantial areas and buildings for alternative uses; because of their historical and monumental value, these elements in the urban fabric cannot easily be used for productive purposes. Rather it has had a social or service effect, as it has freed the city centre from heavy traffic which had become insupportable in terms of congestion and pollution; in 1992 the customs received, on the original site and before the liberalisation of the circulation of goods in the Community, on average between 500 and 800 trailer-trucks per day.

The relocation of the general warehouses, in contrast, has freed a vast area which because of its strategic location in the bottleneck between the historic city to the north of the rail tracks and the ZAI to the south is of crucial importance for the future of the urban settlement. It is not possible to analyse in detail here the problem of the 're-foundation' of South Verona as a distinctive urban environment and the question of the rationalisation of urban transport between the two parts of the city. These two interconnected issues today constitute a major problem for Verona. The removal of the fruit and vegetable market, that now occupies an area of about 60 000 m^2, is certain to have a large-scale urban impact: similarly, the proposed relocation of the railway goods yard away from the main urban station to the European Quadrant will also have major effects. This would provide a direct link between the Brenner and Turin–Venice lines, releasing the central station from most of its operations relating to goods, and enhancing greatly the operational possibilities of the freight village. The establishment of the freight village at Verona has thus already had a profound effect on the city, confirming the decisive role which the transport system plays in moulding the urban area and its environs: the example of Verona can in many ways be considered representative of Italian and wider European experience in this sense. However, the functioning of some of the services that have been developed in the Quadrante Europa may be questioned in terms of the level of synergy achieved within the complex itself and in terms of the role of intermodality in a context of regional development. The answer is only partly positive as, for the time being, the outcome involves primarily the juxtaposition of activities rather than the establishment of interlinked functions. Nevertheless, the Verona freight village, alongside other Italian examples of the same phenomenon, clearly exemplify the recent concentration of intermodal traffic; they represent outcomes of considerable actual and potential significance for the economy of the local areas concerned and for the country as a whole.

INTERMODAL STRUCTURES: AN INTEGRATED SYSTEM?

Returning to the broad pattern of intermodal terminals in Italy – small, medium or large, freestanding or included in freight villages or other structures – we might wonder whether effective connections between them exist or not. No clear-cut and unequivocal answer is available, but only a partial one. Until very recently, terminals were operationally independent from each other: the variety of services provided by the different intermodal structures produced a substantial degree of specialisation, a changing spatial pattern of flows of combined traffic and identifiable catchment areas of individual terminals, rather than a fully integrated interdependent system.

Recently, to optimise both the distribution of traffic flows and the functioning of the structure as a whole, in terms of major national and international flows, some terminals are networked: this means that unit loads moving to or from a variety of terminals are consolidated at a pivot terminal (and therefore handled by its equipment) and transferred directly into railway trucks to form a new unit load. Usually, the shuttles that link up to the pivot terminal operate on a fixed timetable and travel whether all the trucks are filled up or not. This is the case with the shuttle between Turin and Antwerp, for example, that receives in Italy sets of load units from Udine and Bologna. This is also the case with the terminal at Busto Arsizio, near Milan, that receives several shuttle trains per day from Germany and transfers the load units as they arrive to form a daily shuttle train directed towards Pomezia (south of Rome).

It is anticipated that in the not-too-distant future a fully interactive system typical of a true network will ultimately be achieved. A strong and consistent stimulus in this direction will be the ever-closer integration of the economy of the EU. Estimates suggest that combined traffic within the EU will increase three-fold between 1995 and 2005. The increase in goods traffic by road between Northern and Central Europe and Italy is estimated at between 4 and 6% annually, very conservative values if one takes into consideration the opening of the Channel Tunnel (with four daily train shuttles planned between Milan and London) and the opening of Eastern Europe. This will create serious problems because the national Italian road and motorway networks, as well as the European motorway network as a whole, will not be able to sustain such an increase a few years hence.

This brings us again to the problem of an adequate integrated infrastructure, especially as far as rail transport is concerned, to serve combined traffic. The main obstacle on a European scale is how to overcome the different national philosophies determining the strategies of the authorities which manage the rail networks, with a view to an overall European perspective and a Europe-wide supply, together with the need to reconcile the strategies of European companies involved in the operation of combined traffic. Just as important, however, is the financing of the necessary infrastructures and equipment on a community level.

To this difficulty we can add others, partly typical of the Italian situation, partly in common with other European partners. Infrastructural problems include technical incompatibilities between rail networks, the congested Brenner pass route, the expansion of the Verona–Bologna line, and the insufficient number and variety of railway wagons. Management problems involve unsatisfactory delivery times, the security of goods, the capacity of the terminals, and the still inadequate spirit of collaboration between public and private operators. The enhancement of the competitiveness of combined traffic in comparison with road transport lies in overcoming these difficulties.

CONCLUSIONS

Transport policy, and intermodal transport policy in particular, fundamentally concerns infrastructure. However, the efficiency of transport operations and systems depend on the availability and diffusion of information regarding the services and upon the phases within the transport cycle which, in turn, form part of the logistics of enterprise and the logistics of transport in particular. The decisive role which information plays is derived from the particular nature of transport itself. Maximum rationalisation of the supply of transport in relation to meeting demand is essential. Rationalisation needs to involve, first, the meticulous planning of loads and transport operations thereby reducing the costs of middlemen and allowing established and efficient connections between operators, consignors and transporters; and, second, a context of increasing market knowledge realised by databanks, freight exchanges and commodity exchanges. There is substantial and increasing experience at national and European levels, of the first of these requirements. The second, however, has involved numerous difficulties, although it is interesting to note that some databanks and freight exchanges need to be established and managed in specific locations, most appropriately within the intermodal structures and especially the freight villages.

In this context it is argued that the establishment and development of an integrated network of inland load centres, and the hierarchical differentiation between the various terminals within such a network, lie largely in the hands of commercial institutions. In the Italian case, analysis of the present situation shows that the integrated structuring of this hierarchical network has scarcely yet begun and, moreover, is meeting numerous problems in taking off. The main difficulty probably arises from the infrastructural overprovision that brings the various enterprises into competition, along with their reluctance to reveal information concerning supply/demand ratios and operating strategies. Only a hierarchically structured network functioning efficiently enough to permit logistics, and especially transport, to perform to their full potential can ultimately contribute effectively to the organisation and rational modification of spatial patterns.

REFERENCES AND FURTHER READING

Assointerporti (1991), 'Interporti di II livello: cosi il Piano Quinquennale', *Assointerporti News* 1(2–3), 4.

Bologna, S. (1992), 'Sui nuovi metodi per muovere le merci', *Politica ed Economia* 9, 22–26.

Brugiotti, R. (1992), 'Il trasporto combinato per la mobilità europea', *Vie e Trasporti* 585, 13–20.

CEMAT (1994), *Società italiana per il trasporto combinato strada-rotaia* (Roma: Effestudio).

Chiovelli, A., Dondolini, S. and Vitelli, M. (1993), 'Piano generale dei trasporti e interporti. Problematiche connesse con l'attuazione della legge 240/90, capo I per quanto concerne gli interporti di II livello', *Assointerporti News*, 4.

Conti, S. and Julien, A. (eds) (1991), *Miti e realtà del modello italiano. Letture sull'economia periferica* (Bologna: Patron).

CRESME (1993), 'L'inefficienza della gestione del territorio', *Vie e trasporti* 592, 36–43.

Dicken, P. and Lloyd, P.E. (eds) (1990), *Theoretical perspectives in economic geography* (New York: Harper & Row).

European Conference of Ministers of Transport (1993), *Terminology on combined transport* (Paris: ECMT), 1–25.

Ferrovie dello Stato (1993), *L'offerta merci. Servizi, orari, notizie utili sul trasporto merci per ferrovia* (Roma: FS area merci).

Fossati, R. (1992), 'Il trasporto combinato sarà protagonista della rete paneuropea', *Vie e Trasporti* 585, 47–48.

Gazzetta Ufficiale delle Comunità Europea (GUCE), L. 305, 10 Dicembre 1993.

Gertler, M.S. (1992), 'Flexibility revisited: districts, nation-states, and the faces of production', *Transactions of the Institute of British Geographers* 17(3), 259–78.

Hayuth, Y. (1982), 'Intermodal transportation and the hinterland concept', *Tijdschrift voor Economische en Sociale Geografie* 73, 13–21.

Hayuth, Y. (1988), 'Changes on the waterfront: a model-based approach', in Hoyle, B.S., Pinder, D.A. and Husain, M.S. (eds), *Revitalising the waterfront* (London: Belhaven), 52–64.

Hayuth, Y. (1992), 'Multimodal freight transport', in Hoyle, B.S. and Knowles, R.D. (eds), *Modern transport geography* (London: Belhaven), 199–214.

Holtgen, D. (1994), *Intermodal freight centres in Europe* (London: Windborne International).

Hoyle, B.S. (1988), 'Development dynamics at the port–city interface', in Hoyle, B.S., Pinder, D.A. and Husain, M.S. (eds) *Revitalising the waterfront* (London: Belhaven), 3–19.

Knudsen, D. (1994), 'Introduction to the special issue: flexible manufacturing', *Growth and Change* 25(2), 135–43.

McKenzie, D.R., North, M.C. and Smith, D.S. (1989), *Intermodal transportation: the whole story* (Omaha: Simmons).

Pyke, F., Becattini, G. and Sengenberger, W. (eds) (1990), *Industrial districts and inter-firm cooperation* (Geneva: International Institute for Labour Studies).

Rizzo, G. (1980), 'Il trasporto ferroviario delle merci e l'intermodalità', *L'Amministrazione Ferroviaria*, suppl. n. 7.

Rizzo, G. and Robiglio, C. (1990), 'Verona: innovazione e cambiamento', in Borlenghi, E. (ed.), *Città e industria verso gli anni novanta* (Torino: Fondazione G. Agnelli), 97–138.

Russo Frattasi, G.G. (1992), 'La logistica nei trasporti', in Boario, M., De Martini, M. and Di Meo, E. (eds), *Manuale di logistica* (Torino: UTET, vol. 3), 189–288.

Saronni, G. (1993), 'Le FS e il trasporto merci', *Vie e Trasporti* 592, 2–13.

Slack, B. (1990), 'Intermodal transportation in North America and the development of inland load centers', *Professional Geographer* 42, 72–73.

Storper, M. and Scott, A.Y. (1992), *Pathways to industrialization and regional development* (London: Routledge).

Vallega, A. (1984), *Unitizzazione e ciclo di trasporto* (Savona: Camera di Commercio).

Vallega, A. (1985), *Dai porti al sistema portuale* (Savona: Camera di Commercio).

Vallega, A. (1992), *The changing waterfront in coastal area management* (Milan: Franco Angeli).

Part V

PLANNING STRATEGIES AND IMPLICATIONS FOR THE FUTURE

15 Lessons from Massachusetts Coastal Zone Management's Designated Port Area Program: The Fore River Shipyard Re-use Project

LAUREL RAFFERTY
Massachusetts Coastal Zone Management Office, Boston, USA

Over the past quarter century, several major studies have called for better federal planning and coordination of the ports of the United States. These study-findings reflect the status of federal port planning in the USA: unlike in Canada and some other countries around the world, there is no national port policy, or little at least that is readily discernible; there is no central federal agency with financial and administrative review authority over port operations, and no national port development strategy. While a multitude of federal actions (discrete special-purpose subsidies and the rulings of numerous agencies) influence ports, these actions are uncoordinated, frequently leading to unintended, and sometimes detrimental, effects.

Contrary to the repeated study-finding that more central coordination is needed for ports in the USA, the movement in this country has been in the opposite direction. The federal *regulatory* role in port development has been shrinking over the past two decades, the result of the broader national trend in economic deregulation. The strong tradition in the USA of local port autonomy has fostered and reinforced the decentralisation movement. Though the federal government took a one-time stab at promoting port regionalism in the 1970s, it little altered the thriving tradition of local autonomy. By their own design, states and local government have primary responsibility for port administration and finance. This is the assessment of Hershman and Kory: in sum, federal port policy in the USA has been in retrenchment in the 1980s. This is only part of the assessment, however. These authors note the irony of a second trend: that while economic and port development regulations have been in retreat, environmental regulations have been expanding (see note 1); and that while these regulations have had a significant impact on port development in ways not uniformly welcomed by the port community, they have become a vehicle for at least a weak

Cityports, Coastal Zones and Regional Change. Edited by Brian Hoyle.
© 1996 John Wiley & Sons Ltd.

form of federal port policy long sought after in some quarters. The Coastal Zone Management Act, in particular, plays or has the potential to play this role (Hershman and Kory, 1988).

An important element of the federal environmental regulatory arsenal, the Coastal Zone Management Act (CZMA) was adopted in 1972. Voluntary and incentive-based, the CZMA Programme provides for state and municipally developed plans to preserve and enhance coastal resources, both economic and environmental. Accordingly, a key national CZMA goal is the achievement of plans for the wise use of land and water, the promotion of wise or environmentally sound coastal development. Such a goal is sufficiently overarching to encompass a policy to preserve and promote maritime industry.

The avenue CZMA creates for federal port planning is limited in important ways. National CZMA goals are also broad enough to provide for a range of policies potentially in conflict with the fostering of maritime industrial development, policies such as the promotion of recreational activities, public access, and non-industrial, water-dependent uses, and the protection of natural resources. Unmandated, individual state CZM plans may address national goals in a manner that explicitly prevents such potential conflict, but need not do so. State plans need not be consistent with each other at a regional or national level in how they meet CZMA goals.

Advocates of decentralisation suggest that under such a scheme each state can be an experiment in how best to achieve a shared goal – an experiment from which others may learn. The Massachusetts Coastal Zone Management Plan and its Designated Port Area policy is an example of a state plan that specifically resolves, at the state level, potential conflict among port and other land and water uses. A close look at the Massachusetts experience in centralised state planning of its ports offers lessons for other states and demonstrates writ small the issues that will emerge writ large were a comparable policy applied at the national level.

In this chapter the application of the Massachusetts CZM port policy at a specific waterfront location is examined. The site is the former Quincy Fore River Shipyard, now called the Fore River Staging Area.

ISSUES AND OBJECTIVES: STATE AND LOCAL

In 1978, the Massachusetts Coastal Zone Management Program was the first on the east coast to receive federal approval. To implement the Program, the MCZM office was established as part of the Executive Office of Environmental Affairs (Figure 15.1), the lead environmental agency of the state. Financing for the office and programme implementation came through federal funds, one of the key incentives of the CZMA, and was matched 1:1 by the state.

Following a so-called networking model, the MCZM Program is embodied in 27 policies reflecting the mandates of all state agencies which address the preservation and enhancement of coastal resources. The Program also includes the regulations and programmes of these agencies (MCZM, 1991b). One of the

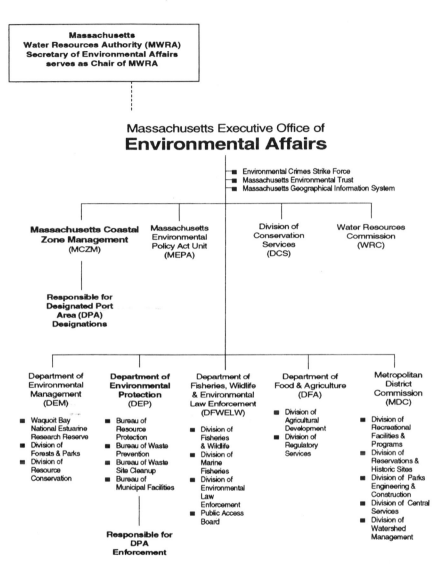

Figure 15.1. The administrative structure of the Massachusetts Executive Office of Environmental Affairs.

27 MCZM policies is directed at preserving and protecting maritime industry and is the basis for the state's Designated Port Area (DPA) Program. This Program establishes the priority of the protection of maritime industrial development over other MCZM policies within designated areas. The DPA policy and programme reflect a mandate of the state Department of Environmental Protection and is enforceable through the regulations of this Department.

The Massachusetts Designated Port Area Program was developed in response to regional changes in land and water use to prevent the conversion of waterfront areas suitable for maritime industry to non-port uses. Twelve designations were established in 1978 (Figure 15.2). In a state with a strong 'home rule' tradition, the DPA Program is one of a very few regional growth management programmes that is legally binding on a municipality. The Program provides at once for state water-dependent industrial zoning and relaxed wetlands protection standards within these same zones, implementing an environmental as well as an economic agenda. It promotes the continued industrial use of areas where industry is able to capitalise on prior infrastructural investments, thereby avoiding the waste of previous investments and the need for redundant new investment. Promoting industrial activity where the environmental impacts of industry have already occurred, it is intended to minimise their spread to previously undeveloped areas.

Methods for implementing and enforcing this policy include combinations of 'carrots and sticks' that have evolved over time to improve the effectiveness of the Program. When first developed, regulations were strong, prohibiting uses that would pre-empt maritime industry. The regulations were limited in area of impact, applying to the water only. Incentives included the promise of preferential treatment in grant funding for DPA projects. In the mid-1980s, regulations were extended to cover the land, when the state's regulatory jurisdiction was expanded to former tideland areas that had been filled to create developable space. When experience demonstrated that regulatory changes were needed to enhance the viability of maritime industry, allowable uses were broadened. They encompassed general industrial activities that would either support maritime development or make only temporary use of a site without altering its character. Limited commercial activity became permissible, if part of a maritime industrial park.

The DPA Program has raised issues with communities with land subject to its requirements. Since the economic downturn of the late 1980s, grant funding for DPA projects was not available at anticipated levels, leaving communities with insufficient wherewithal on their own to induce significant levels of maritime industrial development or maintain essential infrastructure of underutilised areas. Communities have confronted difficult alternatives. If maritime industry did not materialise within designated areas, a community would bear the cost of preserving the land for its intended use. If a community developed a local site plan that was inconsistent with state DPA regulations, a development impasse would result, for required state licences would not be granted for uses unallowable under state requirements. To address the encroachment into DPAs of residential uses and other activities incompatible with industrial revitalisation, a community might require relief from DPA controls. In 1993, at the outset of the Fore River development project, regulations to de-designate or modify DPA boundaries were under development, the criteria for such changes not yet established.

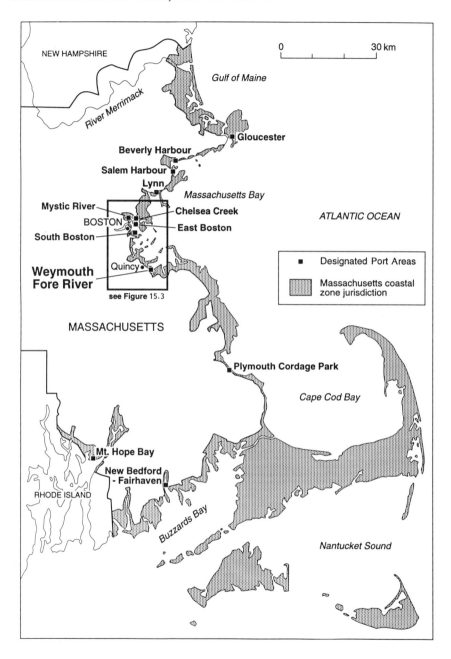

Figure 15.2. Massachusetts, USA: Designated Port Areas

The case of the Fore River Shipyard

Located just south of Boston, Quincy, Massachusetts, is a suburban coastal community, known for naval shipbuilding for over a century. When its major shipyard closed in the mid-1980s, the city lost its primary manufacturing base. Scrambling to adapt to the new economic realities, the community has sought not only to attract new jobs and tax revenues, but also a new image. It has found itself in competition with other communities in the region similarly suffering the abandonment of their traditional industries. Like many, the community desires the benefits of participation in clean, growing economic sectors. It sees its future in the heritage tourist industry. In the face of a vision of the community as a 'City of Museums' lies the presently underutilised, large-scale former shipyard site, falling within a so-called Designated Port Area. This designation, backed up with rigorous regulations, is intended to preserve its use for industrial activity dependent on a waterfront location. This chapter presents an account of the development of a re-use plan for this site that meets both state and local objectives, and links the idea of the changing cityport with the concepts of coastal zone management.

The project site under discussion here is the former General Dynamics Fore River Shipyard, a 180-acre (73 hectare) property located at the southern edge of Boston Harbour, at the mouth of the Fore River. It falls principally in the City of Quincy and extends into the neighbouring Town of Braintree. The Town of Weymouth, located across the river, shares the harbour area (Figure 15.3). The shipyard served as the bread-and-butter industry of its region, providing 30 000 jobs at its peak and contributing significantly to the area's growth. Consistently representing the state-of-the-art in shipbuilding, setting records in production speed in both World Wars, the shipyard generated much hometown pride. As with most ports-of-old, it was integral to community life.

Shortly after its closure, the shipyard was purchased by the Massachusetts Water Resources Authority (MWRA), a state agency responsible for meeting the water and sewage disposal needs of much of the metropolitan Boston region. Charged with cleaning up Boston Harbour, where untreated sewage previously was disposed, the MWRA is using portions of the shipyard site as a staging area for construction associated with its clean-up activities, as well as for a sludge pelletising plant. With the completion of construction activities in 1996, sections of the site not needed for the pelletising plant will become available for redevelopment. With its new, albeit temporary, function, the site is now called the Fore River Staging Area (FRSA) (Figures 15.4 to 15.6).

At the MWRA's initiative, a Development Committee was formed to determine the best re-use of the site. The Committee consisted of representatives of Quincy, Braintree, Weymouth, and MCZM, as well as the MWRA. The Committee selected a consultant team led by Lane, Frenchman and Associates to prepare a development plan. A final development plan was completed in 1994. The Fore River Staging Area Development Plan (1994) meets state and local

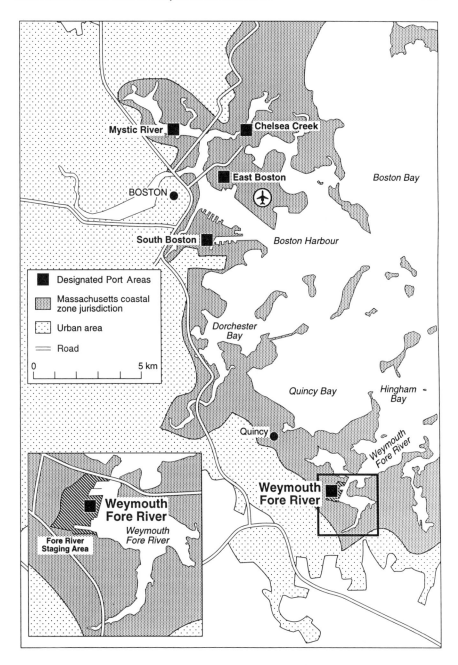

Figure 15.3. Boston, Quincy and the location of the Fore River Staging Area

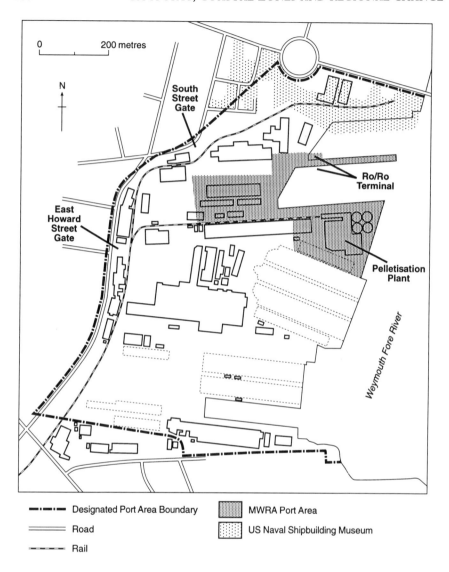

Figure 15.4. The Fore River Staging Area

objectives, the planning process having been designed to ensure this outcome. The planning process included representatives of all stakeholders as principal decision-makers. The state agency responsible for coastal zone management was represented at the table, able to influence the plan's compliance with state policy as the plan was developed. A consensus on a final recommended plan was assured because stakeholders first reached a consensus on key elements of the

Figure 15.5. The Fore River Staging Area, Quincy, Boston, Mass., USA. (Courtesy of Karen-Jayne Dodge, Regina Villa Associates, Massachusetts Water Resources Authority, Boston, Mass., USA)

process: planning goals and principles, alternative development scenarios to be considered, and criteria for evaluating these scenarios.

In achieving agreed-upon goals, the plan represents a model for port development strategies (Figure 15.4). Four key elements are involved. First, the plan maintains the site's capacity to serve as a port. It proposes the revival of shipbuilding and introduction of a maritime technology centre focused on the advancement of ocean engineering and construction, including new methods of shipbuilding. Second, the plan preserves the integrity of the site. The plan calls for a comprehensive treatment of the site, integrating all proposed uses. It preserves both its large-scale industrial capacity and its historic assets. Continued ownership by a single state agency is recommended to ensure that development and implementation take place in accordance with the plan. Third, the plan enhances public use of the site. A tourism centre will commemorate the site's maritime past, present, and future. Tourist activities reflecting this theme will include a naval shipbuilding museum, interpretive facilities at the marine technology centre, and a ferry gateway to regional and local coastal heritage resources. Fourth, the plan encourages a balanced mix of uses. Proposed uses are in traditional manufacturing, high technology, and tourist industries, providing for diversification of the economy and diversification of employment opportunities.

Figure 15.6. The Fore River Staging Area, Quincy, Boston, Mass., USA. (Courtesy of Kevin A. Kirwan, Massachusetts Water Resources Authority, Boston, Mass., USA)

The Fore River project raised issues, particularly concerning the DPA status of the project site, that the plan development process would answer: Would there be a demand for maritime industry at the Fore River site when harbour clean-up activities were completed? Would such activity be consistent with the community's vision? Would the physical attributes of the site be suitable for this activity? Would such activity be economically feasible?

REACHING A SOLUTION

Initial objectives

At the outset of the planning process, none of the bodies represented on the Development Committee had specific long-range comprehensive plans for the site. The initial objectives of these stakeholders indicated the potential for consensus but also for state/local, intra-state, and intra-local conflict. The MWRA had three objectives: to recapture its investment in the property in the interest of its ratepayers; to accommodate an expanded pellet plant operation; and, eventually to dispose of the property. As a state agency, it recognised its obligation to comply with state regulations and maintain good relations with the communities affected by its operations. The City of Quincy had a limited plan for a US Shipbuilding Museum on a 10-acre (4-hectare) corner of the site. All three community members shared a primary concern: minimising the negative impact of the MWRA use of the site. They were acutely aware that they might not share equally in redevelopment impacts, negative or positive: development increasing revenues for the host community might create increased truck and rail impacts for abutting towns.

MCZM's interest was in compliance with its DPA policy and successful port revitalisation. Having no studies to recommend any particular maritime industry as a feasible anchor use, it had numerous concerns: that certain industries would be prematurely written off for lack of sufficient information on their viability; and that the closure of the shipyard would be interpreted as conclusive evidence regarding the prospects for revival of this use.

Site constraints and opportunities

Given its size, waterfront location and infrastructure, the site is recognised as an important resource presenting a development opportunity, though with constraints to overcome. Prior to MWRA take-over, the site contained 78 buildings with a total of approximately two million square feet (186 000 m^2) of space. The vestiges of shipbuilding remained: there are six shipbuilding basins and four outfitting piers as well as a variety of cranes and shipbuilding machinery. Basins include five dry docks and one wet basin; outfitting piers range in length from 560 to 880 ft (171 to 268 metres) and in width from 23 to 43 ft (7 to 13 metres). The MWRA has made some capital investment in the site that was significant to its attracting maritime industry, including the installation of new piers and state-of-the-art roll-on/roll-off facilities. The site has fair regional land access by highway, excellent water access for both movement of freight and a commuter ferry, and is served on-site by the Fore River Railroad. While there is off-site rail access, lines have many street-level crossings, and increased use would affect local street traffic.

In addition to the requirements of its Designated Port Area status, other state regulations affect the site. Existing studies have substantiated that the site is contaminated with hazardous materials, requiring state-regulated remediation. Because of its role in the two World Wars, it is a nationally significant historical landmark, subject to regulations governing historic preservation.

With regard to the pattern of use of the area, maritime industrial activities continue to dominate the general coastal vicinity. These uses have been dynamic, with emerging industries now sometimes replacing traditional manufacturing. More recent residential development has encroached upon the area. Commercial strip development flanks the site's northern edge.

Indicators of the economic health of the surrounding region present a mixed picture. While the population has been stable over the past 20 years, the region's labour force participation has grown significantly. During the 1980s, employment trends fluctuated dramatically, except in manufacturing, which experienced significant decreases in jobs throughout the decade. The closure of the shipyard contributed significantly to this decline. Substantial employment growth enjoyed in all other sectors during most of the 1980s was undermined by substantial losses at the turn of the decade. Industrial clusters of the region include health care; knowledge creation; information technology; financial services; defence, instrument and other manufacturing; and tourism. These clusters reflect the region's high educational attainment levels, which well exceed national levels.

The planning process

The planning process was traditional in format. Five re-use scenarios were identified (Table 15.1). These were derived from an analysis of existing conditions as well as from shared project goals and principles identified during the planning process. The feasibility of each scenario was analysed and the results evaluated in the light of agreed criteria (Table 15.2). Public workshops were held at key junctures in the process to solicit community feedback.

Within 12 months, a consensus emerged. The Development Committee recommended that the site be developed as a centre for marine technology and tourism. Overall, this scenario best met the criteria, with an excellent match in terms of compatibility with site features and consistency with regulations, the highest economic returns, and the fastest market absorption rate, good job generation and tax impacts, and low rail and truck impacts.

The recommended scenario: a centre for technology and tourism

The recommended scenario provides for four catalyst uses that are inter-related in a way intended to support the goals of the plan. These uses include a marine technology centre and associated technology transfer park, a tourism centre, and shipbuilding or another major industrial use. The technology and tourism uses are intended to create a campus of activities on the northern half of the site. The southern half will house heavy and light water-dependent industry, retaining the more traditional industrial character of its shipyard use. The two halves will have separate entrances and the northern campus-like environment, which abuts commercial and residential neighbourhoods, will serve as a transition area, buffering the impacts of the heavier industry and changing the image the site presents to the community. The scenario combines uses from two strong economic sectors of the region: it provides for the expansion of a high-technology industry that is marine-based, and couples this with activities of the burgeoning heritage tourism industry.

The marine technology centre is for the development and advancement of ocean engineering and construction, including the development of new methods of shipbuilding, marine structures of all types, and instrumentation and control systems. It is intended to host an industry genuinely reliant on a waterfront location, and an emerging maritime industry with a future, but related to the historic use of the site. Part of the purpose of the centre is to develop innovative

Table 15.1. Assessment of scenarios

Measures	Scenario 1: incremental use	Scenario 2: traditional redevelopment	Scenario 3: anchor projects	Scenario 4: technology and tourism	Scenario 5A: revitalized port	Scenario 5B: revitalized shipbuilding
Physical compatibility	Good	Fair	Good	Excellent	Poor	Excellent
Total development costs	$31M	$26M	$36M	$34M	$40M	$42M
marketability (absorption)	18 yrs	16 yrs	13 yrs	12 yrs	14 yrs	14 yrs
development index (NPV)	$4.3M	$7.2M	$9.9M	$10.9M	($4.7M)	$10.7M
Jobs added (on-site)	2760	3380	3830	3350	2440	2400
municipal revenue impact	$3.04M	$4.25M	$4.34M	$3.92M	$2.37M	$2.34M
Auto impact (w\mitigation)	low	moderate	moderate	moderate	very low	very low
rail/truck impact	low	low	low	low	high	moderate
DPA compatibility	Good (potential)	Poor	Poor/Fair	Good	Excellent	Excellent
MHC compatibility	Good	Poor	Fair	Excellent	Poor	Good

Source: *Scenarios for Development of the Fore River Staging Area* (September 21, 1993) and *Feasibility Report: Development Scenarios* (January 25, 1994), prepared by the consultant team led by Lane, Frenchman and Associates.

Table 15.2. Scenarios, goals and feasibility criteria

SCENARIOS: The conceptual elements comprising each scenario were first developed. Four such elements were identified by the consultant team:

- Increment of development and disposition: The five identified scenarios ranged in the approach taken on the size of the smallest development unit from individual building spaces to parcels to the entire site. The approach of a given scenario significantly affected all other scenario features such as infrastructure requirements and marketing.
- Mix and location of uses: A consistent vocabulary of uses was considered for all scenarios but with a different emphasis in each. Options ranged from an ad hoc mixture to a special mixed use district to a marine industrial park.
- Critical infrastructure improvements: Access to and within the site was the key issue. It was understood that road access decisions would affect site use and development patterns of each scenario and that broader land and water access needs would depend on other scenario elements.
- Site development and marketing entity: Options included: a non-profit; MWRA; public/private partnership; state port authority; and some combination of the foregoing.

GOALS: The overarching goal was to maximize benefits, while minimizing negative impacts. Specific supporting goals included:

- Enhancing public use of property
- Maintaining site's capacity to serve as a port
- Providing reasonable economic returns to MWRA
- Preserving the physical integrity of the site
- Encouraging a balanced mix of uses

SCENARIO FEASIBILITY CRITERIA: Scenarios were evaluated in terms of five criteria:

- Compatibility with the physical characteristics of the site
- Economic feasibility, including market absorption rate, costs and returns
- Employment and municipal revenue impacts
- Transportation impacts
- Regulatory consistency with state DPA, historic preservation, and hazardous waste regulations

Source: *Scenarios for Development of the Fore River Staging Area* (September 21, 1993) and *Feasibility Report: Development Scenarios* (January 25, 1994), prepared by the consultant team led by Lane, Frenchman and Associates.

technology ensuring the continued global competitiveness of a renewed US commercial shipbuilding industry. It is also to provide for diversification of the economy, generating high-wage sustainable jobs in the service sectors. These considerations have a strong practical as well as thematic basis.

The scenario has a tourism theme intended to commemorate both the marine past of the site and the water-based industry of the present and future. It envisions a synergistic relation between port uses and tourism and capitalises on this connection. It includes interpretive facilities at the marine technology centre and educational programmes that will focus on the activities of this centre and elsewhere on the site. In addition, the scenario provides for the preservation of significant historic structures and development of the US Naval Shipbuilding Museum, including the mooring of a Second World War ship that was built at the site. The scenario also proposes a National Park Service visitor centre to serve as a southern gateway to key regional and local coastal heritage resources. These latter include the Harbour Islands, which played an important role in the defence of the nation from 1634 through to the 1990s.

In the course of plan development, shipbuilding became economically feasible. The US Congress enacted the National Shipbuilding Act, making capital available for ship financing and shipyard modernisation; $230m in loan guarantees were committed to the Fore River Staging Area for these purposes. The restoration of shipbuilding will allow exploitation of the skills of an existing labour force as well as of the existing infrastructure. Technological advances in the contemporary shipbuilding industry, however, will necessitate some labour retraining and infrastructure modernisation. The new federal loan guarantees cover the latter. The plan recommends continued ownership of the site by the MWRA, at least through its complete development. Continuity of ownership under a single public entity is critical to the plan's implementation and will ensure a greater long-term value of the site as a real-estate asset for the MWRA.

LESSONS

It is worthwhile to review pivotal elements of the path to a fast consensus. The make-up of the Development Committee was designed to ensure both representation of a range of relevant interests and decisions satisfactory to these interests. This was a design that worked in practice. Planning goals, the alternative development scenarios to be considered, and scenario feasibility criteria were the outcome of decisions made by the Development Committee. Agreement on these elements of the planning process assured agreement on a development plan that would adequately meet both local and state objectives. A look at the identified goals and feasibility criteria, noted above, readily reveals how the process assured that the recommended plan would incorporate economically and physically feasible maritime industry in a context consistent with the community's vision.

MCZM's direct participation on the Development Committee allowed it to influence the plan's compliance with DPA policy at the plan's making when this was most easily achieved. This influence took a variety of forms. MCZM provided interpretive guidance on state policy, technical information on the needs of the maritime industry, real-world models for proposed scenario elements, and reviews of draft planning documents. The following examples illustrate some of its more significant input. To evaluate the costs of each scenario fairly, the planning consultant identified a baseline infrastructure that all proposed development programmes would need. Serving a dual purpose, this device also identified necessary front-end development costs. While an intriguing conceptual approach, development of a baseline infrastructure could significantly affect the potential for maritime industry, if not constructed with sensitivity to the varying and specialised needs of different industry types. MCZM pushed for identification of a baseline infrastructure that would create parcel configurations not inadvertently exclusive of certain maritime activities. Of particular concern was the impact of the main access road defining the size and configuration of both the buffer area and the remainder of the site. It was clear that the larger the buffer area, the more constrained the area for heavy industry; but other unknown constraints could result. To ensure a sufficient understanding of specialised site needs, MCZM recommended more interviews with a range of port users.

MCZM, vigilant about comments affecting the acceptability of maritime industry, brought to the Development Committee's attention remarks made at public workshops in favour of heavy industry: that heavy as well as light industry was acceptable, if the industry had a future; that the concern about such industries as shipbuilding stemmed more from the fear about their long-term prospects, and less from their noxious impacts. Responding to MCZM's input, the Committee agreed to project goals that did not preclude shipbuilding or other heavy maritime industries. Alternative development scenarios, constructed to be consistent with this Committee decision about goals, provided for heavy as well as light maritime-industrial uses. MCZM was able to learn first-hand the unintended obstacles DPA regulations could present to development options. This was valuable information. While the Development Plan was being formulated, MCZM was revising the DPA Program. These revisions were shaped by MCZM's participation in the Fore River Staging Area Plan development process.

Exercising discipline, project participants did not fall victim to endless scenario development, analyses and evaluation that jeopardises the budget and schedule of many a planning project. The consultant team had generated a limited number of the most feasible scenario components, components that could be mixed and matched in variations on a theme to create the most acceptable overall plan. When shipbuilding was demonstrated to be the winning element of a port revitalisation strategy, it was incorporated in the marine technology and tourism scenario. Discerning the elegance of this newly formed alternative, the alternative which became the recommended scenario, MCZM concentrated its energies on determining compliance issues of this scenario and how to resolve them.

The project profited from extremely strong project management. A representative of the MWRA, the property-owner, took responsibility for this function. This arrangement was highly unusual and was perceived by some as akin to letting the fox guard the chicken coop. Despite organisational affiliation, the MWRA manager proved good at balancing the need to satisfy the MWRA Board as well as the communities and MCZM and received strong support from the planning consultant in this effort. Paid for by the MWRA, communities, and a state grant, the consultant gave his professional judgement even when contrary to the perceived interests of individual committee members, effectively marshalled compelling evidence to support these judgements, and offered creative solutions to accommodate all sides. The consultant took on battles on behalf of the project at times when there was no internal committee dispute; e.g. to preserve historic assets and prevent design and site planning impacts that would be detrimental to development potential. The members of the consultant team demonstrated a genuine appreciation for the site's port and historic assets, and were particularly knowledgeable about the site, having overseen a prior university study of the project area. This university project, while providing substantial information on existing conditions, had also served to float trial balloons on potential re-uses and sparked creative thinking about new possibilities.

Success of the project is considered to depend on the development of catalyst uses: shipbuilding, a major industrial use, the National Park Service programme, and the marine technology centre. However, human catalysts, not catalyst uses, will matter most. As Hoyle and Pinder have written: behavioural, not physical, geography is determinative of port vitality (Hoyle and Pinder, 1992). Behavioural geography has already mattered. Plan implementation is under way because of the continuing efforts of the project consultant and Development Committee as well as other government agencies, whose support they have enlisted. The commitment of federal loan guarantees and a marketing effort have produced strong interest by shipbuilders in the site. The National Park Service has shown the Fore River site as one of the potential gateways to the Boston Harbour Islands in a recently released study. A draft prospectus for the centre for marine engineering and industry has been developed in application for federal funding. Efforts are under way to involve a consortium of partners in the centre, including two world-class marine science programmes in the region, the Massachusetts Institute of Technology and the Woods Hole Oceanographic Institute, as well as the Massachusetts Maritime Academy, all of which have expressed an interest in participating in the centre.

CONCLUSIONS

Without the state restrictions imposed under the DPA Program, the former Fore River Shipyard site could very likely have bowed to the development pressures of the 1980s that led to the conversion of other Quincy waterfront locations to non-water-dependent and non-industrial uses, such as condominiums and marinas. The state port area designation maintained the availability of the Fore River Shipyard site for an important public, albeit temporary, water-dependent industrial use – a staging area for construction activity associated with the clean-up of Boston Harbour. It is not clear that there would have been an alternative location suitable for this purpose. If development pressures had not converted the site to non-port uses in the 1980s, these pressures surely would have prevailed with disposal of the property by the MWRA in the 1990s, were it not for the site's DPA status. If converted to non-port uses, a scarce and shrinking waterfront resource would have been irreversibly lost, and a domino effect on surrounding maritime industrial uses would most likely have been inevitable.

Local controls alone cannot be depended on to protect a state-wide interest in the preservation of suitable areas for maritime industry. Local zoning of the Fore River site in Quincy allows for water-dependent industry as well as other types of mixed-use development potentially inconsistent with DPA regulations. Quincy's zoning is typical of municipal interest in a level of use flexibility that significantly erodes the ability to preserve a working waterfront. Without state guidance, communities are apt to create such mixed-use waterfront zones because they have ample evidence that existing incentives are inadequate to promote maritime industry, while having precious little information to demonstrate that such industries will prove viable and significantly revenue-generating. They are also apt to have conflicting use policies reflecting the difficulty they have reaching a consensus on these matters and the lack of sufficient technical assistance on dispute-resolution techniques. Finally, they may prohibit maritime industry through zoning to change the industrial character of their waterfronts, without appreciating how aggressive mitigation programmes can buffer impacts perceived to be undesirable.

DPA status was sufficient to preserve the Fore River site for maritime industry, but not to promote such development in this location. While state law does not mandate consistency between local waterfront zoning and state DPA regulations, state DPA requirements are, nonetheless, sufficient to block non-water-dependent, non-industrial projects that are put forward based on local zoning. Local zoning, however, can block maritime development in DPAs. For maritime development to go forward, local zoning has to permit this activity. In the case of the Fore River site, the communities' zoning met this threshold requirement, but a number of other conditions had to pertain for maritime development to occur in accordance with DPA objectives and for de facto land-banking to be forestalled.

The Fore River case study amply demonstrates other conditions that were important factors in the promotion of maritime industrial development:

(1) the use of a carefully constructed plan development process with consensus as an objective;

(2) MCZM's direct participation in the planning process, which helped shape development alternatives meeting state as well as local objectives, thereby avoiding the development *impasse* that would have occurred with overlapping inconsistent jurisdiction controls;

(3) the selection of a development plan in accordance with agreed criteria responsive to concerns about the long-term viability of a proposed use, levels of job and tax revenue generation, the mitigation of negative impacts, community image, and compliance with state regulations;

(4) the changes made by MCZM to the DPA regulations to enhance the economic feasibility of maritime industry, based on the Fore River as well as other real-world experience;

(5) development of a shared vision for the Fore River site that, rather than seeing maritime industrial and non-port uses as mutually exclusive, saw the potential for a synergistic relation between them; and

(6) the availability of public subsidy for port uses, that is, the powerful incentive of the $230m in federal loan guarantees for ship financing and site modernisation for shipbuilding.

The Fore River Staging Area development project serves as an exemplary model of DPA master planning. But even so, based on its experience with this site and other DPAs, MCZM has realised that the promotion of maritime industry requires much more than was in place at the time this project was initiated – more than the preservation of suitable waterfront areas, more than increasing the flexibility of regulations to foster the economic viability of such uses, and more than the other factors noted above. Promotion of maritime industry requires other types of far-reaching changes, most significantly a *state-wide* port development strategy. The Fore River development project illustrates this in a number of ways: for example, in the difficulty it subsequently experienced in finding available sources of capital financing to fund such things as the baseline infrastructure programme, a key element of the implementation strategy; and in the difficulty of coordinating with other port development projects to best fit into a broader port planning framework and to complement, not to compete in dysfunctional ways with, the activities of these ports.

The far-reaching changes the Fore River project has revealed as wanting are being addressed. A new state-wide port development strategy is under development (Governor's Commission, 1994). This strategy includes a $300m capital investment programme, funded through a state bond bill. If passed by the state legislature in the form recommended, the Bill will fund improvements to port infrastructure, including dredging projects, a state-wide doublestack train network, public port facilities, economic development and port planning, as well as a host of other activities. The strategy also includes the creation of the Seaport Advisory Council, a focal point for coordinated decision-making about the state's ports. Consisting of representatives of the state port authority, transportation, environmental, economic, and finance agencies, port communities, and port users, the Council will decide the state's allocation of capital investment in specific port projects as well as address state-wide policy, research and planning needs. MCZM is a key player in shaping this state-wide port development strategy and, in doing so, is making good on the intent of its DPA programme to realise the full potential of the economic resource the state's ports represent. MCZM was successful in getting capital funding earmarked primarily for DPAs, thus fulfilling its original intent to give preferential treatment to DPAs in the award of grants. Such capital is intended to be available for implementation of the Fore River Development Plan. With this achievement, MCZM is improving how it addresses the goal of the national CZMA, to preserve and enhance economic coastal resources.

While centralised port planning and coordination is in retrenchment at the national level, it is blossoming in Massachusetts. Centralised planning and coordination of port development, that is, the dovetailing of the state-wide port development strategy with a state-wide land-use and environmental programme, the state's DPA programme, is enhancing the viability of maritime development. Both elements of this centralised programme have been found critical to this intent. They may be understood to be direct products of the MCZM Plan and the CZMA, legacies of the once-expanding national environmental agenda.

The MCZM Plan is serving to foster port development precisely because it is giving due consideration to economic as well as environmental resources. The hue and cry against the regional land-use and environmental policy the DPA Program represents is now heard far less, as are the cries of those with a competing interest in potentially conflicting use policies also promoted as MCZM goals. Communities are embracing the DPA Program as the state puts all the pieces in place to a state-wide port development strategy, most significantly, a capital investment programme to maintain and improve port infrastructure and fund port economic development, making maritime industry feasible.

One of the key factors in the success of state-wide port development strategy is the leadership role of the chief executive of the state in calling for and putting forth the state-wide port development strategy as his Seaport Action Plan. Through his leadership, the Governor of the state has effectively communicated to port communities and users and state agencies that he has a serious interest in the port economy.

The plan development process is as equally important as the role of executive leadership to the potential for success of a centralised port plan in Massachusetts. A state-wide port development strategy is evolving that is neither the result of a top-down nor bottom-up approach, but rather, is the result of a genuine partnership, a perhaps overused, but none the less appropriate term. The partnership includes all relevant parties, and each partner has, in effect, sufficient power to feel it is gaining control of the future of its area of concern, not losing, by participating. The state-wide partnership mirrors the partnership that worked so well in the development of the Fore River Staging Area Final Plan. This suggests a good planning practice at work.

The Massachusetts DPA Program, coupled with its state-wide port development strategy, and grounded in the CZM, is a model worthy of consideration for replication nationally and internationally. The model, in its essence, is the use of the partnership concept to arrive at and apply state-wide policy. The state-wide policy clarifies how competing interests and uses can be prioritised in the service of both economic and environmental coastal resources, including both the industrial economic resources and the recreational.

NOTES

1. A number of changes have occurred since the 1980s. There were some positive developments in federal port planning in early 1990: e.g. two important pieces of federal legislation were enacted, the federal Intermodal Transportation Surface Efficiency Act (ITSEA) which mandated the integration of port access plans and intermodal management systems into coordinated state and regional transportation plans; and the National Shipbuilding Act promoting both the conversion of US shipyards to commercial shipbuilding and the development of maritime technology. These are in jeopardy, however, as increased decentralisation is the intent of the significantly changed US Congress of the mid-1990s. While the results are not yet in, elimination of many environmental, as well as economic, regulations has been proposed, including the CZMA.
2. Massachusetts' protection of the public interest in all tidelands is based on law known as the public trust doctrine; as a result, concerns about running foul of the 'taking issue' do not have the same relevance as they may when a municipality uses restrictive zoning to preserve land for water-dependent uses.
3. The jurisdiction of the Massachusetts Port Authority is limited to the Port of Boston.

REFERENCES AND FURTHER READING

Centre for Real Estate Development, Massachusetts Institute of Technology (1992), *Fore River Shipyard Re-Use Study Briefing Book, Existing Conditions and Opportunities*.
Charlier, J.J. (1992), 'The regeneration of old port areas for new port uses', in Hoyle, B.S. and Pinder, D.A. (eds), *European port cities in transition* (London: Belhaven), 137–54.
Code of Massachusetts Regulations:
    301 CMR 9.00 Waterways Regulations.
    301 CMR 23.00 Regulations for the Review and Approval of Municipal Harbor Plans.
    301 CMR 25.00 Designation of Port Areas.
Goodwin, R. (1988), 'Waterfront revitalization: ways to retain maritime industries', in Hershman, M.J. (ed.), *Urban ports and harbour management* (New York: Taylor & Francis), 287–305.
Governor's Commission on Commonwealth (of Massachusetts) Port Development (1994), *Final Report*, October.
Hershman, M.J. and Kory, M. (1988), 'Federal port policy', in Hershman M.J. (ed.), *Urban ports and harbor management* (New York: Taylor & Francis), 99–122.
Hoyle, B.S. and Pinder, D.A. (1992), 'Cities and the sea: change and development in contemporary Europe', in Hoyle, B.S. and Pinder, D.A. (eds), *European port cities in transition* (London: Belhaven), 1–19.
Lane, Frenchman and Associates, Inc. (lead consultant), and Centre Associates, Howard/Stein-Hudson, Parsons Brinkerhoff, and Philip B. Herr and Associates (sub-consultants):

Fore River Staging Area Development Plan Technical Memorandum 1: Precedents for Development (August 1993).

Fore River Staging Area Development Plan Technical Memorandum 2: Issues facing the future of the Fore River Staging Area (August 1993).

Fore River Staging Area Development Plan Technical Memorandum 3: Participation process for planning: participation plan (June 1993); Report on participation (July 1993); Goals and principles for redevelopment (July 1993).

Fore River Staging Area Development Plan Technical Memorandum 4: Feasibility report on development scenarios (July 1994).

Massachusetts Coastal Zone Management/Executive Office of Environmental Affairs (1991a), *MCZM: a comprehensive tool to protect marine resources* (Coastal Brief, No. 8).

Massachusetts Coastal Zone Management/Executive Office of Environmental Affairs (1991b), *EOEA and the coastal programme* (Coastal Brief, No. 10).

Massachusetts Coastal Zone Management Programme (1977), Volumes I and II.

Massachusetts General Laws Chapter 91, *The Public Waterfront Act.*

16 Cityports, Coastal Zones and Sustainable Development

ADALBERTO VALLEGA
Department of Urban, Regional and Landscape Planning, University of Genoa, Italy

DIACHRONIC MODELS

The current changes in cityports, as well as related changes in their regional role, are rooted in social and spatial processes which have been developing at least from the first Industrial Revolution. These processes have passed through various morphogenetic phases, during which the spatial division of labour and a range of consolidated spatial structures collapsed and new structures were built up. Geography has focused upon these processes using stage-based models concerned with the evolution in social organisation, cityports and coastal industrialisation.

Social and spatial changes

From the geographical point of view the stage-based models of the evolution of society in terms of *social and spatial changes* built up by Geddes (1915) and Mumford (1934) maintain their interest for at least two reasons: (i) they focused on the historical evolution of the interactions between technological advance, social organisation and spatial structures, and (ii) they were mainly concerned with the global scale, embracing the developed countries as a whole. According to Geddes' model, nations and regions have passed through two stages: palaeotechnical and neo-technical. Mumford's model led to the identification of three stages: eo-technical, palaeo-technical and neo-technical. Moving from these approaches and bearing in mind the social and spatial changes which have occurred in the second half of this century, a comprehensive model based on the palaeo-, neo- and post-industrial stages, can be designed (Vallega, 1992).

Shifting from a broad and comprehensive view of social and spatial changes to a view specifically concerned with those changes which cityports have undergone, Hoyle (1988) proposed a diachronic model of *cityport evolution* embracing five stages: (i) the primitive cityport, (ii) the expanding cityport, (iii) the modern industrial cityport, (iv) the retreat of the city from the waterfront, (v) the

Cityports, Coastal Zones and Regional Change. Edited by Brian Hoyle.
© 1996 John Wiley & Sons Ltd.

redevelopment of the waterfront. The evolution of the industrial functions of the cityport – the process of *cityport industrialisation* – has been the core subject of a wide spectrum of geographical investigations, many of which have highlighted the clear links between the increasing deadweight tonnage of bulk carriers, on the one hand, and the growing dimensions of coastal manufacturing plants, especially iron and steel industries, oil refineries and power plants, on the other. Vigarié (1981) succeeded in building up a stage-based model in which the evolution of coastal industries was explained through the concept of the Maritime Industrial Development Area (MIDA) and the identification of four generations of MIDAs.

These three approaches – social and spatial changes, cityport evolution, and cityport industrialisation – could be considered jointly with the aim of providing a single view of the evolution of cityports and coastal regions. Although such an attempt could be regarded as methodologically inappropriate it is a useful way of approaching an issue – i.e. the target toward which the regional role of the cityport is moving – which is one of the most complicated aspects of current spatial change. The neo- and post-industrial stages merit close attention.

Neo-industrial cityports and coastal regions

Based on the techniques of converting thermal and mechanical energy into electrical energy, on the use of oil and gas energy sources, on geothermal energy exploitation and, lastly, on the use of nuclear fuels, the neo-industrial society was born in the late 19th century in the United States, spread into Western Europe and Japan from the 1930s onwards and to other parts of the world from the 1950s. The *modern cityport* was its focal spatial structure. In this respect, Hoyle's model (1988) points out that both coastal industrialisation and the conversion of general cargo traffic into containerised flows were the main causes of the decline of links between the city and the modern port which had marked the neo-industrial stage. As a consequence, both the changing coastal industrialisation and the birth and diffusion of unitisation for the transport of manufactured goods were leading spatial processes.

According to Vigarié's stage-based model (1981), the maturity phase of the neo-industrial stage was characterised by two generations of MIDAs. The first MIDA generation developed during the period from the late 1950s to the late 1960s. In Europe, the Botlek terminal at Rotterdam (1958) may be regarded as the initial step towards MIDAs since it brought about a diffusion wave of similar developments which, in a short time, involved many ports facing the North Sea, the Atlantic Ocean and the Mediterranean Sea. These areas were served by port terminals capable of unloading minerals and energy sources carried from increasingly distant mining regions. A second generation of MIDAs developed during the 1970s as efforts to mitigate environmental impacts from iron and steel industries, oil refineries, petrochemical and other raw-material processing plants, as well as the need to promote the location of more labour-intensive manufacturing, emerged in some cityports, especially in the Netherlands.

Unitisation and containerisation

While coastal industrialisation, sustained by the scale economies and subsequent propensity to giantism in carriers and processing plants, was characterised by the building up of MIDAs distant from the seaport and brought about a spatial separation from the cityport, the second separation factor, namely containerisation, was developing. To highlight how and to what extent this process has exerted influence in the cityport and coastal region functions during the maturity of neo-industrial society, the following stage-based model of containerisation may be considered.

(i) Preparation for take off and diffusion (1958–64). During this phase, the first generation of cellular vessels, capable of transporting less than 1000 TEUs (twenty-foot equivalent units), was experimentally used on short sea routes along the coasts of the United States and from their seaports and those of Puerto Rico, as well as on mid-sea routes from the United States to Venezuela. Conventional port structures were used to deal with this new standardisation-inspired transportation tool.

(ii) Take off (1964–70). The second generation of cellular vessels, capable of carrying 1200 to 2000 TEUs, emerged and the containerisation of deep-sea routes was initiated in the Atlantic Ocean. The *Multimodal Transport Operator* (MTO) and *Combined Transport Operator* (CTO) were the protagonists of the progressive transfer of manufactured goods transportation from the break-bulk cargo system to unitisation. Specialised terminals for lo–lo and ro–ro handling systems and standardised superstructures were introduced. As a result, standardisation in maritime transportation and seaports began also to influence land transportation.

(iii) Drive to maturity (1970s). During this decade the third generation of cellular vessels, capable of 2500 to 3000 TEUs, was introduced. The *Through Transport Operator* (TTO), able to serve all the containerised and break-bulk-based phases of the transportation cycle, began to compete with the MTO and CTO. Together with the use of increasingly large carriers this contributed to the building of lo–lo terminals distant from the port, provoking the dissociation between the city and the port. In the meantime the first generation of inland load centres was designed, bringing about the initial step towards a regional network based on the unitised transport cycle.

According to Hoyle's model (1988), the converging consequences from the process involving both general and unitised cargoes, on the one hand, and bulk cargoes, on the other, were the main causes of the changes in the waterfront's functional role. Changes passed through two phases.

(i) The retreat of industrial and general cargo-based facilities from the seaport. During this phase, lasting from the late 1960s to the late 1970s, the separation

between the port and the city, which began during the maturity of the neo-industrial stage, reached its peak.

(ii) The take-off of waterfront redevelopment. Cityports, reacting to the de-location process of port functions, started planning new waterfront functions, based on both sea-oriented and non-sea-oriented facilities.

SUPPORTING CONCEPTS

At this point in the discussion it might be useful to recall that geographical literature and, *sensu lato*, the literature on spatial processes has explained the role of the cities facing economic growth using the concepts of central place, industrial pole and gateway. Christaller's approach (1933) would not have had the importance acquired during the 1960s if Berry (1964) and others had not dealt with the need to describe the diffusion processes of urban structures, which were primarily a consequence of industrial growth. The metaphor of the central place, through which city development and urbanisation were explained, was based on the consideration of commercial functions, jointly with another strict set of tertiary functions – especially transportation and administration – and the distinction between the urban and extra-urban market to which they referred. As a consequence, the theory of central places is to be regarded as an explanation of the tertiary polarisation energy of the city. Hierarchy is the leading concept through which the region can be identified and described: the rank of tertiary functions of the city leads to the identification of the functional rank of the city within a system of cities. The gravitation area associated with the city endowed with the highest functional rank coincides with the extent of the region. The regional organisation is imagined as a hierarchical pyramid consisting of elements (central places) more or less related to each other (relationships) and serving also an extra-urban population (gravitation area).

The spatial structure by which the strong polarisation provoked by the industrial growth was explained was the industrial growth pole, or economic pole, a concept which was forged during the 1960s moving from the Perroux basic approach (1964). The idea of a leading industry, thought of as a top-level spatial structure diffusing effects on a surrounding space and transforming it into a region, reflected the need to put into focus not industry *per se*, but big industry, which was characterising economic evolution in developed countries.

Looking at the geographical literature it appears evident that both the concepts of central place and industrial growth pole were applied mainly to investigate the spatial processes concerned with inland areas. Cityports and coastal industrial areas were not widely involved in this approach although coastal areas were affected by huge urbanisation and MIDAs were the foci of economic growth. On the contrary, the concept of the gateway was forged by Bird (1981) with the role of the cityport in mind. 'Gateway functions are those that link a home region to other regions in the nation state and the nation state to the rest of the world via international transport. Gateways therefore stand in contrast to central places which serve the "land around" – the umland'. These are not two distinct images of a city but two spectra of city functions. 'As a gateway grows larger, it begins to have central place impacts on the home region. As a central place grows larger, it must have its extra-regional links' acquiring gateway functions (Bird, 1981, 196).

SPATIAL IMPLICATIONS OF POST-INDUSTRIAL SOCIETY

The background consisting of both models of social change and spatial processes, as well as the concept of polarisation effects, could be regarded as containing useful tools to deal with processes brought about by the emerging post-industrial stage. To contribute to the building of conceptual tools and models capable of dealing with this unprecedented complexity, we can consider coastal industrialisation and gateway functions which have been developed since the 1980s as a consequence of the evolution of unitised transportation.

Changing coastal industrialisation

According to Vigarié's model (1981), during the 1980s two generations of MIDAs, the third and fourth, emerged. They arose from many factors, among which changes in the international division of labour had a leading role. During the 1980s the tendency to locate manufacturing plants, especially heavy industry, in developing countries, which had been initiated in the previous decade, developed and diffused in many areas, such as South America and South East Asia. The final product of this process was the third generation of MIDAs, which was based on the iron and steel industry, oil refineries, bauxite refineries and other activities.

The developed world, at least the most advanced countries, reacting to this de-location process, moved towards the creation of the fourth generation of MIDAs. Mainly consisting of the transformation of first generation areas, these areas are marked by the decline or conversion of heavy industries, such as iron and steel plants and oil refineries, and the development of finished goods processing plants, high-technology industries and an increasingly large integrated network of tertiary activities.

Hoyle's approach (1988, 10–11) leads to a focus on the profound changes induced in city–port relationships because of the sequence of MIDAs which has emerged from the 1960s. The first and second generations of MIDAs brought about a dissociation between industry and city, which was both *spatial*, since MIDAs were located in places distant from the city, and *functional* since the need for tertiary urban activities from those manufacturing plants was not enough to give shape to strong integration. One should wonder whether current changes in these areas will lead to the establishment of functional links between manufacturing activities and the city. As a preliminary approach, it should be agreed that, the larger the spectrum of finished goods production, high-tech manufacturing processes and tertiary activities of the converted or newly created MIDAs, the greater the prospect of implementing links between these new multiple-function areas and urban growth.

Gateways and unitised transport cycles

A look at the processes capable of influencing the role of cityports in their regional contexts can focus on gateway functions arising from changes in the unitised transportation paying attention to two questions, namely, whether they (i) mitigate the dissociation between the modern seaport and the city and (ii) bring about effects involving the cityport and industrial areas jointly. According to the answers given to these questions, the expected role of the cityport varies. On this subject at least three technological and organisational factors, depending on the role of unitised (especially containerised) transportation, are to be borne in mind: (i) the round-the-world routes, both westbound and eastbound, have been experimented with success and pendulum routes have been introduced; (ii) increasingly large cellular vessels were introduced and, during the 1980s, the 4000 TEUs threshold was passed; (iii) in the early 1990s the 6000 TEUs capacity, namely overpanamax, was regarded as a target of the new generations of these vessels; (iv) advanced computer and telematics systems have given an unexpected possibility not only of optimising the management of the containerised cycle but also of tendering services for extra-cycle operations.

These spatial processes justify considering the gateway functions of the cityport on the basis of changing the unitised, especially containerised, transportation cycle. Bearing in mind that containerisation covers almost the totality of manufactured goods and food traffic, it follows that: (i) load centres are to be regarded as top-level gateways for the inter-regional and international relationships; (ii) the containerised seaport competes with the inland load centre to catalyse gateway functions; (iii) changes in the location of these gateway functions influence the evolution of the spatial division of labour on both inter-regional and international scales.

The gateway role of the seaport, rooted in containerised traffic, emerged in the late 1970s (United States) and early 1980s (Europe and other areas) and has significantly evolved in the 1990s. This is mainly due to the strategies designed by the so-called *global superstars* such as American President Lines, Maersk and Sea-Land, and others such as Nippon Uysen Kaisha and P&O Containers which have played a leading role. Thanks to the powerful information and telematics systems they manage, these operators are able to take care of (i) the final phases of the manufacturing process and (ii) the distribution phase. Acting as *Transport and Distribution Operators* they manage a wide spectrum of extra-transport functions – such as the assembling of components, packaging, labelling and other characteristics of the conventional manufacturing processes – the final result of which is the preparation of the goods for the various sector and area markets. In addition they provide the necessary facilities and services for the distribution of goods to buyers. This step concludes a profound evolutionary process, rich in spatial manifestations, which started with the Through Transport Operator: in its initial phase, the operator managing containerised cycles was successful in serving also conventional transportation and, in the current phase, he cooperates with producers maximising the width of his operational sectors and optimising the scale economies provided by the facilities initially built up only for transportation.

The more the interaction between producers ands so-called global superstars makes progress, the more the load centre acquires also the role of *logistic centre*. From this arise impulses for the location and development of industrial facilities and functions. The final result of this process is the birth of a *multiple-function pole*, basically consisting of facilities and services for multimodal and combined transportation cycles as well as for break-bulk operations, manufacturing processes managed by transport operators and similar processes managed by producers, services for the distribution of manufactured goods, and tertiary activities acting as external economies.

This pole is sustained by a feedback opposite to that which occurred during the neo-industrial stage. In the past the initial input was caused by industrial growth while, in the current phase, the changing organisation of the transport cycle is able to influence industry, even in a deterministic way. In the past, feedback consisted of a series of relationships: (i) the growth of dimensions of coastal processing plants to optimise the scale economies; (ii) the growth of deadweight tonnage in bulk maritime transportation to reduce the raw material costs; (iii) the expansion of industry-serving seaport facilities and, as a consequence, the increase in maritime transportation. During the current phase another chain of relationships emerges: (i) the increasing global productivity of fully or partially container-based transportation cycles; (ii) the stimuli to create logistic centres, including seaports, in places suitable for the optimising of the distribution of manufactured goods; (iii) the growth of external economies in logistic centres attracting the location of the final phases of the production processes and operations preparatory to distribution; and (iv) as a consequence, the growth of containerised traffic and the increase in productivity of the containerised cycle.

Shifting from these spatial processes to the regional role of the cityport, three deductions seem justified.

(i) The model of the fourth generation of MIDAs merits being re-considered in the light of the logistic centre-based strategies of producers and the global strategies of the most advanced transport operators. This could be helpful for visualising the prospects for the conversion of the MIDAs' functional profile, moving from heavy to light manufacturing processes and from a single-feature profile, namely industrial, to a three-feature profile, namely industry-, transportation- and distribution-based.

(ii) The post-industrial stage is characterised by unprecedented and growing integration between transportation and manufacturers which will exert influence in seaport and coastal manufacturing areas.

(iii) The *spatial dissociation* between the cityport and both the seaport and industrial areas is expected to remain, since the containerised cycle, manufacturing processes and logistic functions need to be located away from the urban area.

(iv) Nevertheless the *functional dissociation* between the cityport, on the one hand, and the containerised seaport and industrial area on the other, could decrease due to the strength of growing interaction between manufacturing and tertiary activities.

The revitalised waterfront, the converted MIDA, and the containerised terminal – all of these viewed in connection with the birth and diffusion of logistic centres – are the focus of investigations aimed at building up scenarios on the future cityport. These spatial structures must also be considered in the light of the possible effects of the sustainable development policy adopted by the United Nations Conference on Environment and Development (UNCED, 1992).

THE CITYPORT AND COASTAL REGION AFTER RIO

According to the background which supported the UNCED approach, sustainable development is conceived as the result of policy contextually pursuing (i) the integrity of the ecosystem, (ii) the efficiency of the economy and (iii) social equity, including inter-generation equity. Moving from this concept Agenda 21, Chapter 17, provides principles and guidelines for all kinds of seas, including coastal areas. Although the coastal region and cityport concepts are not included in the conceptual spectrum of Agenda 21, Chapter 17, they have elements from which geography could design those concepts that are consistent with the sustainable development paradigm. In this respect, if the integrity of the ecosystem is considered as the basic goal of sustainable development, the coastal region should coincide with one coastal ecosystem or a set of contiguous ecosystems, in such a way as to optimise their management through a well-tailored decision-making system. As a result, moving seaward three ecosystems acquire interest: (i) the land ecosystem involved by coastal facilities and resource uses, (ii) the fresh–salt water ecosystem and (iii) the marine ecosystem. The continental margin is the main abiotic reference. By this approach coastal management embraces a system of actions with the aim of: (i) providing an integrated use of natural resources guaranteeing the integrity of the ecosystem, (ii) minimising possible conflicts between uses, and (iii) transferring natural and cultural patrimony to future generations.

The building up of a sustainable development-pursuing coastal region presupposes that only uses of natural and cultural resources consistent with the three principles of the basic paradigm – namely, integrity of the ecosystem, economic efficiency and social equity – are developed. The spectrum of coastal uses consistent with sustainable development-referred discriminants is very large. As a result, it should be agreed that the more the coastal region moves towards sustainable development-inspired targets pursuing integrated management of the environment and resources, the greater the possibility of widening the range of coastal functions.

If this statement is accepted, the cityport is expected to develop action on three scales – urban, regional and international – according to the extent to which sustainable development-consistent functions are expanded. An evident feedback links the three scales: the more the cityport develops sustainable development-referred functions, the more it is able to serve as a main reference basis for regional policy and the more it is able to attract attention from the international market of clean technology and emerging tertiary activities.

No detailed framework of functions is presented here, but it is enough to point out that the sustainable cityport is expected to:

(i) implement functions, consisting of research, scenario-building and planning, including management programmes for natural cycles, especially the sea-level variations and subsequent implications;

(ii) develop functions based on the management of biomass and its living resources with research into fishing, aquaculture, the protection of resilience and the biodiversity of food chains;

(iii) expand functions aimed at maximising the use of renewable resources, especially energy sources, and minimising that of non-renewable ones;

(iv) encourage as much as possible technologies aimed at providing comprehensive monitoring of the coastal system, including the ecosystem and coastal uses;

(v) develop long-term programmes for the protected areas in order to stimulate environmentally-sound enjoyment of the natural patrimony;

(vi) develop programmes for the management of the cultural patrimony regarding it as both an economic resource and as an ethics-relevant component of the coastal community.

During the neo-industrial stage the regional role of the cityport was perceived only in terms of catalysing capability – in terms of population, traffic flows and others – so the regional extent was thought of as coincident with that of the gravitation area created by the central place- or industrial pole-based functions. During the emerging post-industrial, sustainable development-influenced, stage the regional role of the cityport should be perceived through a more extended range of elements. Its ability to drive the region towards sustainable development-consistent targets is expected to become more important, at least on two levels.

(i) The *environmental level* since, thanks to the functions it has, the cityport should be in a position to provide management programmes and tools, as well as technical assistance, to guarantee the integrity of the ecosystem, namely dealing with sea-level variations, erosion cycles and other, including human, factors. (ii) The *development level* since a cityport that is keen to develop a driving role – up to the point of acting as a regional capital – is expected to provide a basic set of functions stimulating the sustainable development of its region. Research, education and training, central and local headquarters of inter-governmental organisations and national agencies, offices of non-governmental organisations, financial enterprises, mass media headquarters, and many others, are to be considered as components of this emerging functional level regardless of to what extent – if at all – they participate in sustainable-referred programmes and actions.

The final step of reasoning leads back to the starting point, namely, the role of the functional and spatial components of the cityport and their interaction. Four kinds of structure are involved: the central business district; the redeveloped waterfront; the containerised seaport and, *sensu lato*, the unitised functions of the seaport; and the maritime industrial development area.

The *central business district* is relevant to the prospect of driving the cityport towards a new regional role to the extent that it is able to host decision-making bodies operating in the areas embraced by sustainable development policy. The conventional central business district was mainly concerned with banks, insurance and trading. The new district is expected to include, as its main components, bodies concerned with environmental management, research, mass communication and the management of its cultural heritage.

The *waterfront* functions should include facilities and services centred on the protection of the coastal and ocean ecosystems, the assessment of the sea, including that based on educational interactive methodologies, and the assessment of cultural heritage that is concerned with the marine environment. This should be regarded as an evolution of the social perception of human–ocean relationships, the final target of which could be a cultural background much wider than that provided by the conventional approach to the environment. In this context, the waterfront could acquire high-rank functions, dealing with research, university education, decision-making systems, and others. As far as the hierarchy is concerned, these functions are similar to those ascribable to the central business district.

The *containerised seaport* is developing sustainable development-consistent functions according to the extent to which it: (i) maximises the role of the load centres (logistic platform) as the top-level expression of its gateway role; (ii) does not bring about environmental implications. Finally, the role of the *maritime industrial development area* evolves because of a wide range of factors, which play different roles according to the coastal regions and cityports. The decline of heavy processing plants in the developed world associated with the increase of those of developing and newly industrialised countries, the emerging catalysing role of the containerised seaport also for manufacturing operations, the increasing importance of products sustained by bio-engineering, computer systems and telematics, and others, converge to bring about a very complicated web of inputs and processes, which perhaps cannot be included in a holistic explanatory model.

CONCLUSION

Two scientific needs are becoming more socially important. On the one hand, the need to understand the changes occurring in the coastal areas and their cities, hopefully achieving comprehensive views. On the other hand, the need to build up sustainable development-aimed planning and coastal management programmes. A retroaction links these two needs: (i) the understanding of the evolution of the cityport and its region towards a post-industrial functional profile stimulates sustainability inspired planning; (ii) planning helps to understand processes. In this context three scales are involved:

(i) the *intra-urban scale*, in which some kinds of strategic spatial structures – the central business district, the waterfront, containerised and associated functions, the manufacturing area – are of local interest;
(ii) the *urban scale*, which is concerned with both the role of the city as a whole, i.e. regarded as the final result of the inputs arising from urban functions, and the regional role of the cityport, i.e. in terms of its capability of influencing and driving the region to which it belongs;
(iii) the *regional scale*, relating to the coastal region.

Analyses carried out on these three scales should be integrated. There is no doubt that the more integration is developed the more research acquires scientific justification and social importance.

At present these subject areas are dealt with by many disciplines, social and natural. As far as geography is concerned, many sectors are involved, to various extents within physical, human, economic and regional geography. Integration between these approaches is needed. As a conclusion, the prospects for research are fascinating if two issues are productively tackled. First, an *epistemological issue*, since the integration between disciplines, as well as that between sectors of a single discipline, requires the adoption of a paradigm and theoretical background. Second, an *assessment issue*, which could be pursued by the setting up of an *observatory of cityports and coastal regions* focusing on sustainable development-relevant processes, is needed both for research and planning.

REFERENCES

Berry, B.J.L. (1964), 'Approaches to regional analysis: a synthesis', *Annals of the Association of American Geographers* 54, 2–11.
Bird, J.H. (1981), 'Gateways: slow recognition but irresistible rise', *Tijdschrift voor*

Economische en Sociale Geographie **74** (3), 196–202.

Christaller, W. (1933), *Die zentralen Orte in Süddeutschland*, (Wissenschaftliche Buchgesellschaft). English edition, *Central places in southern Germany* (ed. C.W. Baskin) (Englewood Cliffs: Prentice-Hall).

Geddes, P. (1915), *Cities in evolution, an introduction to the town planning movement and the study of cities* (London: William and Norgale).

Hoyle, B.S. (1988), 'Development dynamics at the port–city interface', in Hoyle, B.S., Pinder, D.A. and Husain, M.S. (eds), *Revitalising the waterfront. International dimensions of dockland redevelopment* (London: Belhaven), 3–19.

Mumford, L. (1934), *Technics and civilization* (New York: Harcourt and Brace).

Perroux, F. (1964), *L'économie du XXe siècle* (Paris: Presses Universitaires de France).

UNCED (1992), *United Nations Conference on Environment and Development* (New York: United Nations).

Vallega, A. (1992), *Changing waterfront in coastal area management* (Milan: Franco Angeli).

Vallega, A. (1996), 'The coastal use structure within the coastal system: a sustainable development-consistent approach', *Journal of Marine Systems* **7**, 95–115.

Vigarié, A. (1981), 'Maritime industrial development areas: structural evolution and implications for regional development', in Hoyle, B.S. and Pinder, D.A. (eds), *Cityport industrialization and regional development: spatial analysis and planning strategies* (Oxford: Pergamon Press), 23–37.

Index

References to Figures are shown in **bold** type

Abidjan, Côte d'Ivoire 162
Acts of Parliament (UK) 22
Actual Environmental Impact (AEI) 63–5
Adriatic Sea 105–19
Aesthetic values 61
Affluent periphery, in Italy 245
Africa, new ports 170–2
    port cities and cityport regions in 160–2
    see also East Africa
African cityports 160–2
African economies, weak state of 175
Aggressive mitigation 290
Agora 141
Air pollution 88, 91–2, 97–8, 100
Air transport 115, 117–18, 146–7
Akashi-Kaikyo Bridge (Japan) 214
Albania 105–19, **108**, 148
Albert Dock, Liverpool (UK) 32–3
Alternative development strategies 281, 287–8
Ancona (Italy) 141
Animal transport 113
Arch bridges 214
Ardrossan (UK) 86, 88, 91, 94
Ashworth, G. 185
Associated British Ports (ABP) 14–16, 30
Association Communautaire de l'Estuaire de la Loire 24
Association Internationale Villes et Ports (AIVP) 3, 176
Athens (Greece) 124, 127, 133
Avon County Council 22
Avon Gorge 12
Avon, R. 11
Avonmouth (UK) 12, 14, 21, 22
    coastal zone 20

Balanced mix of port uses 281
Balkans 105–19, **107**

Barcelona (Spain) 141
Barry (UK) 15
Baseline infrastructure 288, 291
Behavioural geography 289
Belfast (UK) 86, 88, 92, 98
Belgrade-Bar railway 106
Bell Lines 20
Berry, B.J.L. 298
Bi-modal transport systems 105
Bi-polar conurbation 240
Bird, J.H. 298–9
Birmingham (UK) 94
Bitumen 94, 97–8
Black Sea 115–16
Bologna (Italy) 257–8
Boston (USA) 5
    Harbour 278, 289
Bottlenecks 109–12
Boyer, M. 236–7
Braudel, F. 141
Brindisi (Italy) 255
Bristol (UK), Chamber of Commerce 20
    city centre dock redevelopment 16
    City Council 12, 22
    port 9–25
        evolution 11
        improvement scheme 12
        ratepayers 13
    Britain 131–2
        Greek commercial links with 123
    British Coal 16
    British electricity industry
        privatization 21
        energy policy 15
        trade 123, 131–2
    British Transport Docks Board 14
    British Petroleum (BP) 86, 89, 92, 94, 97–8
    Buchanan (Liberia) 172
    Building industry 72
        materials, quality of 72–3

Bulgaria   111, 115–16
Bulk import cargoes   9, 12
   handling terminals   16–17
Burmah   97
Burton, Sir R.   167, 217
Business parks   14
   traffic   222

Cable-stayed bridges   214
Canary Wharf   201–2, 206
Car imports   19–20
Cardiff (UK)   15, 23
   Cardiff Bay Barrage   23
Casual labour   15
Catalyst uses   284
Cattedown, Plymouth (UK)   63, 65–82, 67
   demographic structure   73
   economic structure,   71–2
   land-use evolution   65–9
   oil trade   69
   planning proposals   78–80, 79
   residential structure   72–3
Central Business District   304
Central and eastern Europe   105–19
Central Place Theory   298
Centralized planning, of port
   development   274, 292
Centre for marine engineering and
   industry, Boston (USA)   289
Centre for marine technology and tourism,
   Boston (USA)   284
Centro Internazionale Città d'Acqua,
   Venice (Italy)   2–3
Chamber of Commerce, Bristol (UK)   20
Channel Tunnel (UK-France)   189, 215–
   19, 228, 230, 231, 267–8
Chemical complex, Bristol (UK)   12
Christaller, W.   298
Citizens Advisory Committee   81
Cityports   61, 84
   in Africa   160–2
   evolution of   295–6
   industrialization   295–6
Cityport-region relationships   1–2
Clark University (USA)   65
Clyde, R. and estuary   3, 15, 24, 49-57,
   50
Clyde Valley Regional Plan   52, 54
Coal   99
   imports of   17
   industry   15
   coal-fired power stations   16

coal terminal (Bristol)   19–20
Coastal Planning: Planning Guidance Note
   (PPG20)   22
Coastal heritage resources   281, 287
Coastal zones   61, 82
   defences   22
Coastal zone management   20–5, 91,
   140–2, 305
   cultural heritage   140–1
   environmental problems   140–1
   planning dimensions   140
   strategies   186
Coastal Zone Management Act (CZMA)
   (USA)   274
Cognitive distance   80
Collisions   95
Colonial port cities   160
Combined transport, in Italy   249–70, 257
   operator   295
Commercial land use   68
Communication networks   65, 69
Community image   291
Community Tolerance Limit (CTL)   63,
   65, 68–70
Commuter ferry   283
Confederation of British Industry (CBI)   20
Conflict situations, in coastal zones   61–5,
   71–2, 82
Consensus   280–90
Conservation, urban   173
Conservative Government (UK)   14
Constraints on fixed link
   construction   213–18
Container transport   105
   containerization   99, 101, 295
Copenhagen (Denmark)   223, 225, 228
   airport   225
Côte d'Ivoire   160, 175
Counterurbanization   244
Cultural functional links   62, 72
Cultural images   190–1
Currants   121, 129, 133–4
   trade   129–33

Dakar (Sénégal)   162, 176
Danube, R.   111
Decentralization   273
Decision making process   63
Defence industry   24
Deindustrialization   84–92, 100–1
Demand-led economies   249
Denmark   223

Derelict land   77–8
Deregulation   14, 273
Designated Port Areas, Massachusetts
  (USA)   277, 278
Development alternatives   291
Development Area status   24
Development plans   62
Devonport (Plymouth) (UK)   68
Didcot power station (UK)   19
Disinvestment   83–92
Diffuse industrialization   243
Discharges   95
Dispersants   95
Dispute-resolution techniques   290
Distance friction   224
Diversification   196
  of economy   281
Docks, dock gates (Bristol, UK)   13
Dockside land, extensive areas of   19
Dover (UK)   188–94, 189, 205
Duarte Barbosa   167
Durrës (Albania)   109–17
Duty-free sales   222

East Africa, coastal zone   169
  transport services   163
East Denmark   223–4
East Kent ports   219
Economic characteristics   62
Economic coastal resources   292
Economic development   39
  diversification   188, 193–4
  efficiency   81
  functional links   62, 72
  leakages   117
  restructuring process   62
Ecosystems   89
Educational attainment levels   284
Effluents   92
Electric cars   215
Electricity   52–3, 55
  generation   99
  industry, privatization of   15
Electrification   112
Ellesmere Port (UK)   86, 88, 92, 94, 97
Entrepreneurial management   14
Environment   62–75
  aesthetic characteristics of   75
  physical characteristics   67, 75
  social characteristics   75
  and tourism   186
  urban   81

Environmental change   65
  constraints, on estuary development   24
  damage   83
  factors, in Kenya coastal zone   164–5
  gain   84, 88, 91–101
  groups   62–3, 65
  impact statements   91
  impacts   70, 276
    tolerance of   63–5
  legislation   22
  loss   98–101
  monitoring   91
  perceptions   67–8
  planning agenda (UK)   20
  quality   62–3, 77, 81
  regulations   83, 273–4
England, south-east   97
English Channel ferries   222
English Tourist Board   186
Employment   99, 101, 235
  impacts   219
  losses   94
  opportunities, diversification of   281
  trends   283–4
Ermoupolis (Greece)   124
Esso   86, 88, 92, 94
Estuaries   24
  planning, European context of   23–5
Estuary Management Group (Severn)
  (UK)   24
Estuary Zone Management Plan (EZMP)
  (Severn) (UK)   24
ESTURIALES   24
Euclidean distance   79
Europe, central and eastern   105–19
  western   84, 86, 105
European context of estuary planning
  23–5
European Union   116, 222, 225
  combined transport   253, 254
Eurostar trains   223, 228
Eurotunnel   5
Exhaust gases   215

Federal port planning (USA)   273–4
Fehmarn Belt fixed link (Denmark/
  Germany)   225–6
Ferries   109, 111
Fertiliser plant   12, 19
First Corporate Shipping Ltd   16–17
First-generation MIDAs   245
Fishing   71–2

Fitzpatrick, J.   34
Fixed links   5, 219, 222–3
Flexible production   250
Flexible work practices   19
Food processing   72
Fordism   250
Fore River Staging Area, Massachusetts
   (USA)   273–92, 279, 280, 281, 282
   Development Plan   278–9
Foreign aid   113–18
Foreign direct investment   105, 117
Fort Jesus, Mombasa (Kenya)   168
Forth, R. and estuary   15
   road bridge   217
France   132
Freight villages   256–9
French trade   132
Free ports/free trade zones   105, 115,
   147
   at Liverpool (UK)   33
   at Thessaloniki (Greece)   148
Friction of distance   224

Gas-fired power station   21
Gateways, gateway functions   1, 106, 118,
   222
   and regional development   245
   theory   298–9
Gaussian curve   64
Geddes, P.   295
Gedi (Kenya)   167
General Dynamics Fore River Shipyard
   (USA)   278
Genoa (Italy)   141, 153–4
Getz, D.   184
Gilbert, D.C.   187
Girder bridges   214
Glasgow (UK)   11, 49–54
Global competitiveness   285
Gracey, T.   170
Great Belt ferry services (Denmark)   223
   fixed link   223, 225
   East Bridge   214
Great Britain, fixed links   229
Greece   106, 109, 112, 115
Greek cities   141
   government policy   131–2
   War of Independence (1821–30)   123,
   124
Growth poles   245

Handley, J.   35–6

Hamburg (Germany)   106, 228
Harbour Revision Order (HRO)   16–17
Hartlepool (UK)   15, 194–9, 197, 205
Harvey, D.   198
Health problems   22
Heavy industry   288
Heavy metal pollution   22
Helsingør (Denmark)   221
Henry, I.   185
Her Majesty's Inspectorate of
   Pollution   89
Heritage, and tourism   184–206, 278,
   284
Hershman, M.J.   273–4
Heseltine, M.   34
Heysham (UK)   86, 88, 98–9
Heavy goods vehicles (HGV)   99–101
High technology growth   14
   industry   284
High tides, Venice (Italy)   241
Hinterlands   240
   expansion and competition   224–8
   population   16
Historic assets   287
   preservation of   283
Historical factors, in Kenya coastal
   zone   165–9
Honshu-Shikoku Bridge (Japan)   223
House of Commons Public Accounts
   Committee   22
Housing   72, 77–8
Hoyle, B.S.   293–5, 299
Humber Bridge (UK)   214, 218
Hunterston (UK)   49, 51–6, 54, 55

ICDs   2, 174
Image promotion   199
   reconstruction   190
Imports   94, 98–9
Industry   129
   and demographic balance   237
   and port development, in Africa   172
Industry-city dissociation   301
Industrial capacity   281
   clusters   284
   decentralization, in Veneto (Italy)   243
   development   61, 63, 70–2, 78
      attitudes towards   70
   expansion   240–2
   growth poles   298
   land use   68
   redevelopment   80

restructuring 83
tourism 196
Industrialization, and regional
    development 240–1
diffuse 243
Information meetings 81
Infrastructures 173
Ingham, K. 167
Inland container depots (ICDs) 2, 174
Innovative technology 287
Integrated environmental strategies 25
Interest groups 63
Intermediacy 222
Intermodal freight trains 228
Intermodal Transportation Surface
    Efficiency Act (ISTEA) (USA) 293
Intermodalism 2, 242, 250–2
    within the EU 253–5
    in Italy 255–6
International straits 213
Intrusive parking 77
Inward investment 118
Ionian Islands (Greece) 122–3, 129
Ireland 19
Islam 117
Isle of Grain (UK) 86, 89, 92, 94, 97, 99–
    100, 99
Italy 109, 115–18
    late modernization of 245
    road/rail intermodality 255–6

Japan 224–5
    inter-island fixed links 226
    waterfront redevelopment 141
Job creation 186, 198, 284
    and tax revenue generation 291
Journeys-to-work 93, 98, 100–1

Kattegat ferry services (Denmark) 223
Kent 190, 194
    County Council 100
Kenya 162
    coastal zone 162–74, 164
Kilifi (Kenya) 165
Kilindini Harbour, Mombasa
    (Kenya) 166, 169
Kilwa (Tanzania) 167
Kingsnorth (UK) 86, 89, 97, 99
KONVER programme (EU) 24
Kory, M. 273–4

Labour 121, 129, 132

Labour Party (UK) 12, 15
Lagos (Nigeria) 162
Lamu (Kenya) 165–6, 170–1
Land use 72, 77, 79
Land-use change 77
    competition 61
Land banking 290
Landlocked states 106
Land route concentration 222
Large-scale industrialization 244–5
Lead and zinc smelter 12, 19
Leisure facilities 77
    policy 185
    traffic 222
Legal notice 81
    system 63
    tolerance limit (LTL) 63–4, 70
Legislation 63
Le Shuttle trains 223
LIFE programme (EU) 24
Light industry 98, 101
Lille (France) 228
Liverpool (UK) 11, 31
    airport 34
    North Docks 33
    South Docks 34
    see also Mersey
Load unitization 250
Local communities 88, 92, 97, 100–1
    economic development and marketing
        strategies 19
    economic policy 185, 193
    environment 62–3, 65, 69, 75, 77
    desirable improvements 76–7
    government review 22
    plans 63, 78, 80
    port autonomy (USA) 273–4
    zoning 290
Logistics 105
    logistic centre 301
Loire, R. (France) 24
Lomé (Togo) 172
London (UK) 201, 203, 228
    Docklands (UK) 199–205, 200, 203, 204
London Docklands Development
    Corporation (LDDC) 199
'Loony left' 14
Lubricant production 97

M4 motorway and corridor (UK) 14,
    20
Macedonia 106, 109, 111–12, 115–16

Malindi (Kenya)   165–7, 218
Malmö (Sweden)   225
Manchester (UK)   94
  Ship Canal   30–1
  Trafford Park Euroterminal (UK)   228
Manda, Manda Bay (Kenya)   166, 170–2,
  171
Maputo (Mozambique)   162
Marghera (Italy)   237
  see also Venice, Porto Marghera
Marine technology centre   284–5
Maritime Industrial Development Areas
  (MIDAs)   23, 83, 296
Maritime industry   276, 288
  development of   274–6, 290
  preserving and promoting   274–5, 291
Maritime technology centre   281
Maritime transport, technological change
  in   71
Marittima port zone, Venice (Italy)   235
Market absorption rate   284
Market areas   224
Marseille (France)   141
Massachusetts Coastal Zone Management
  programme   274–5
  policy   274
  Designated Port Area programme   276,
  277
  Water Resource Authority   278
Matthiessen, C.W.   225
Medieval cityports   11
Mediterranean, eastern   123
Medway (UK)   15
Mersey Basin Campaign   34
Mersey Docks and Harbour Board   30, 31
  later Company   32–3
Mersey estuary   3, 27–47
  cityport   31
  seaport system   30–1
  zone   28–30
Mersey estuary management plan   29, 37–
  44
  policy proposals   40–4
  vision statement   40–2
Mersey R., proposed tidal barrage   34–5
Merseyside Development Corporation   33
Messina Strait (Italy)   214
Mestre (Italy)   237
Metallurgy   50, 55
Meteorological conditions   97, 100
Metropolitan development, in Italy   245
MIDAs   5, 294–7

Middle East   115
Migration   111, 129, 133
Milan (Italy)   244
Milford Haven (UK)   86, 88, 89, 90, 92,
  94–7
Ministry of Defence (MOD) (UK)   71
Mitigation   291
Mixed-use development   290
MOD   71
Mogadishu (Somalia)   162
Molesworth, Sir G.   170
Mombasa (Kenya)   4, 159, 160–79, 166
Monsoons   165
Mordaunt, T.   16
Motorways   14, 16, 23
Multimodalism   2, 219, 250
  multimodal transport operator   297
Multiple function pole   301
Mumford, L.   295
Muncipal docks, Bristol (UK)   13
Murayama, Y.   224

Naples (Italy)   141, 258–9
National Dock Labour Scheme (UK)   15–
  16, 33
National dock strike (UK)   13
National Power   16–17
National Rivers Authority (UK)   89
National Shipbuilding Act (USA)   287, 293
National Union of Mineworkers (NUM)
  (UK)   17
Natural gas   99
Naval dockyards   68, 71
Navigation, commercial   44
  and harbour conservation   16
Navigational clearance   213
Neo-Fordism   249–50
Neo-industrial society   296
Neo-Romanticism   70
New River Gorge Bridge (USA)   214
'new world disorder'   106, 118
Newport (UK)   19, 23
Nigeria   160, 175
North Sea–Black Sea waterway   106
Nottingham (UK)   94
Nouadhibou (Mauritania)   172
Nuclear power   16
Nyika   164–5

Oil products   69, 72
  distribution   94, 98, 100
Oil refining industry   84–9, 85

Oil traffic 241–2
Ord, D. 16
Øresund fixed link (Denmark/Sweden) 215, 225
  ro-ro ferry 221
Øresundstad (Denmark/Sweden) 225
Outer Hebrides (UK) 218, 228
Overconcentration, of port activity 176
Owen, Capt W.F.W. 170

Padua dry port (Italy) 242
Page, S. 203
Parliamentary Select Committee (UK) 18
Parma (Italy) 257–8
Patras (Greece) 4, 121–36, **122, 123, 128**
Pearce, D. 184
Pembrokeshire Coast National Park (UK) 88
Perceived Environmental Impact (PEI) 64–5, 70–1, 75–6
Perceptions 62–3, 70, 75, 77–8, 80–1
Peripherality 240
Periplus of the Erythraean Sea 166
Perroux, F. 298
Petroleum 109
Pilot Actions for Combined Transport (PACT) (EU) 255
Pinder, D.A. 63
Pipelines 94
Piraeus (Greece) 124, 127
Plan implementation 287
Planning 63, 68
  goals 281, 287
  process 278–9, 282, 284, 287, 291
  proposals 62, 80, 81
  system 62–3, 77–9
Plymouth (UK) 4, 61–82, **66**
  economy 71
  local plan 78
  Planning Department 81
  waterfront 65, 67
Polarizing effects 237
Policy framework 39
Political factors, in East Africa 169
Poll Tax capping 14
Pollution 32, 70, 77
  control 39, 43, 95
Poon, A. 185
Population growth, in Patras 124–9, 133
Port activities 72
  concentration 176
  in East Africa 168–9

development strategy, in Massachusetts (USA) 291–2
  in USA 273, 281
  in greater Venice (Italy) 237
  growth 83
Port-city interface, in East Africa 160
  relationships 1–3
Port-hinterland links 2–3
Port-industry development, and environmental impacts 236–7
  and regional development 244–6
Port management 95–7
Port and regional transport chain 242–4
Port regionalism, in USA 273
Port revitalization strategy 288
Port Talbot (UK) 15
Port users 288
Portbury (UK) 23
Porto Marghera, Venice (Italy) 5, 235, 237, **238, 239,** 241–6
Ports Bill (UK, 1991) 15
Ports, privatization 9, 14–20, 25
Portuguese, in East Africa 167–8
Post-Fordism 249
PowerGen 16
Prentice, R. 191
Private Members' Bill 12
Privatization, of ports 9, 14–20, 25
Product distribution 93
Project management 289
Property values 73
Protection, of Venice (Italy) 241
Public expectations 62
  hearing 81
  inquiry 17
  perceptions 78–9
Public–private partnerships 19
Ptolemy's *Geographia* 166

Quadrante Europa, Verona (Italy) 259–67, **260–65**
  spatial impacts 263–7
Quality of life 70, 75, 82
Questionnaire survey, in Plymouth 63
  hypotheses 74–5
  sampling 73–4
Quincy, Massachusetts (USA) 278
  Fore River Shipyard 273–94

Railways and rail transport 20, 53
  Albania 112–14, **113**
  Balkans 111–12, 115–18

Railways and rail transport (*cont.*)
  East Africa   168–9
  Europe, central and eastern   105–6
  Greece   127
  impacts   284
Real estate market   73
Recession   13
Recreation   39
Recreational facilities   77, 79
  lack of   70
Redevelopment, of ports   112
  impacts   283
Regional electricity suppliers   16
Regional growth management
  programmes   276
Regional land access   283
Regional restructuring   136–57
Residential change   74
  development   78
  land use   68
Restructuring, in heavy industries and
  employment   242
Re-use scenarios   284
Revitalization   62, 81
  see waterfronts
Rezende's map of Mombasa   168
Rhapta (East Africa)   166
Rhine-Main-Danube canal   106
Rias, on East African coast   165
Richards Bay (South Africa)   170
Risk, perception of   65, 69
Rivalta-Scrivia (Italy)   257–9
Roads   53
  traffic   69
  transport   93–5, 98, 105, 109, 112–18
Rødbyhavn-Puttgarden ferry (Denmark/
  Germany)   228
Roman cities   141
Ro-ro vehicle ferries   213, 218, 228
Rotterdam (Netherlands)   83, 296
Royal Portbury Dock, Bristol (UK)   18
Rural settlements   127–9, 134

Safety   75–9
  constraints   218
Sailortowns   62
Saldanha Bay (South Africa)   172
San Leonardo oil terminal, Venice
  (Italy)   241
San Pedro (Côte d'Ivoire)   172
Sanitary conditions   80
Sarandë (Albania)   109, 110, 111

SCOSLA (Standing Conference of
  Severnside Local Authorities)   23
Sea transport   88, 94, 99, 101, 105–19
Seabank Power   21
Seaforth   31
Seagrasses   95
Second industrial zone, Porto Marghera
  (Italy)   235
Second Lake Washington Bridge
  (USA)   215
Second port debate (Kenya)   170–2
Second Severn Crossing (UK)   20, 22–3
Seikan Tunnel (Japan)   215
Semantic differential scale   75
Serbia   105–6, 111–12, 116–17
Service sector   14
Severn estuary (UK)   3, 9–25
  barrage   23
  bridge   20, 223
  ports   10
Severnside local authorities   22
  superport   12
  regional strategy   12–13
  report (1971)   14, 23
Shaw, G.   182, 185
Shell   86, 94, 98
Shëngjin (Albania)   109, 111, 141
Shinkansen   225
Shipbuilding   235, 278, 281, 287
Short sea crossings   213, 218–9
Simon, D.   161–2, 175
Sinclair, M.T.   203, 263
Sites of Special Scientific Interest
  (SSSIs)   22, 32
Skye Bridge (UK)   217
Slave trade, in 19th-century East
  Africa   168
Social construction of tourism   184
Social justice   81
Society's tolerance limits   63
Soil pollution   93
Solvent recovery   98
Sorkin, M.   184
South Africa   175
South Scandinavia   225
  fixed links   227
South Wales   12, 20, 24
  coal industry   9, 17
  ports   9
Southampton (UK)   16
Spain   19
Spatial deconcentration processes   244

Spheres of influence 118
Spillages 89, 95–7, 101
Stage-based historical models 293–4
Stakeholders 280, 282
Standing Conference of Severnside Local
  Authorities (SCOSLA) 23
'Stations of amplification' 65
Stockholm (Sweden) 225, 228
Strait of Gibraltar (Gibraltar/Morocco)
  214
Strategic constraints 216–17
Strategy Plan, Avonmouth/Severnside
  (UK) 23
Structural determinants of tourism 185
Structured workshops 81
Subjective environmental impact 64
Suez Canal 168
'Sunbelt city' 14
Suspension bridges 214
Sustainable coastal region 302–3
  cityport 303–4
  development 5–6, 24, 40, 43
  principle 302
Symbolic images 184
Syra (see Ermoupolis) 124
Swahili people 162

Tagus, R. 24
Tanzania 160
Tate Gallery North 33
Tax impacts 284
Technological change, in maritime
  transport 61
Technical constraints 213–16
Technopole 246
Tees, R. 15
Teesport (UK) 86, 88, 92, 94
Teesside (UK) 194–9
  Development Corporation 195, 196–7
Telecommunications 113, 116
Tema (Ghana) 170, 172
Territorial planning 82
Thamesport (UK) 99
Thessaloniki (Greece) 4, 106, 109, 111,
  133, 137–57
  port 150
  Roman and Byzantine periods 145
  site 143
  waterfront 148, 154
Thomas, B. 198
Through transport operators 297
Tidal range 12

Tideland areas 276
Tilbury (UK) 15–16
Tirana (Albania) 115–17
Tokyo (Japan) 225
Tolerance, to environmental impacts 63–
  5, 64
Tourism, tourist industry 4, 52, 56
  and environment 186
  centre (Boston, USA) 281
  impacts 186
  in Kenya 173
  policy 184–8, 206
  production and consumption of 182
  social construction of 184
  structural determinants of 185
  theme (USA) 285
  theoretical perspectives 182–7
  in Venice 243
Tourist gaze 184, 193, 198, 206
Trade Unions 15
Traffic 69–71, 77–8
  diversion and generation of 222–4
  shadow 225
  pollution 92–3, 98, 100
Train à grande vitesse (TGV) 223, 228
  TGV Belge 223
Training/retraining 113, 115
Trans-European motorway 115
Transferability 224
Transport Act (1981) (UK) 14
  cost barrier 213, 218
  demand 217–18
  and distribution operators 300–11
  geography, of Europe, new 253
  hubs 222
Transshipment terminals 101
Trieste (Italy) 106, 141
Truck and rail impacts 283
Trust ports 15
Turin (Italy) 257–8
Turn-round times, for ships 19

Unemployment rates 14
Unilever 30
Union Railway 223
United Kingdom (UK) 100
  east coast ports 9, 13
  environmental monitoring 91
  oil refinery restructuring 84–101
  planning structure 22
United Nations Conference on
  Environment and Development

(UNCED)   300
United Nations sanctions   111–12, 116
United Molasses   19
United States, interstate highway
   system   217
Unitised transportation   300
Urban conservation, in Kenya   173
Urban development, at Patras
   (Greece)   124, 127
Urban Development Corporations   15,
   188, 197
Urban hierarchy   225
   landscapes   68
   and rural tourism   185
   tourism   196, 205
   waterfronts   61
Urry, J.   182, 184–5, 188
Use flexibility   290
User tolls   217

Vallega, A.   45, 141–2, 295
Values, local   70
Vasco da Gama   167
Vehicle repair   72
Ventilation   215
Via Egnatia   109, 144–5
Vigarié, A.   296, 299
Vision   246, 288
   vision statement, Mersey estuary
   (UK)   40–2
Visual appearance, of buildings   73, 75, 80
Visual pollution   88, 92, 98, 100
Veneto region   236, 240, 245–6
Venice (Italy)   5, 106, 141, 236–48, 239
   conurbation   237
   first industrial zone   235, 246
   General Plan (1965)   235, 241
   lagoon   235, 238
   protection of   241
   second industrial zone   ...
   third industrial zone   240
   and regional development   244–5
Verona (Italy)   5, 249–70

freight village   259–67, 260–5
   spatial impacts   263–7
   see also Quadrante Europa
Vlorë (Albania)   109, 111, 115
Voogd, H.   185

Water access   283
Water-dependent industry   284, 290
   zoning   276
Water-dependent uses, in port areas   274
Waterfronts   61, 65, 68, 71–2, 77, 78, 81,
   181, 198-9, 201, 206
   historical development of   138–9
   preservation of suitable   291
   retreat from   160
   revitalization   2–3, 62, 78, 84, 91, 98–
   101, 139, 188
   at Thessaloniki (Greece)   151–4
   tourism   181–2
   working   290
Water pollution   88, 92, 95–8, 100
Water quality   39, 43
Water sport   72
Wear, R.   24
Welsh Development Agency   24
West Africa   11
West Denmark   223–4
Western Development Partnership   19, 24
Western Europe   84, 86, 105
Wetlands   83, 89, 97
   protection standards, in USA   276
White Cliffs Experience, Dover
   (UK)   191–4, 191, 192
Wildfowl   32, 95
Williams, A.   182, 185
Woodspring (UK)   22

Yugoslavia (former), free zone, at
   Thessaloniki (Greece)   148

Zaire   175
Zanzibar   168–9